AF360885

Diversity of Coral-Associated Fauna

Diversity of Coral-Associated Fauna

Editor

Simone Montano

MDPI • Basel • Beijing • Wuhan • Barcelona • Belgrade • Manchester • Tokyo • Cluj • Tianjin

Editor
Simone Montano
University of Milano
Italy

Editorial Office
MDPI
St. Alban-Anlage 66
4052 Basel, Switzerland

This is a reprint of articles from the Special Issue published online in the open access journal *Diversity* (ISSN 1424-2818) (available at: https://www.mdpi.com/journal/diversity/special_issues/coral_associated_fauna).

For citation purposes, cite each article independently as indicated on the article page online and as indicated below:

LastName, A.A.; LastName, B.B.; LastName, C.C. Article Title. *Journal Name* **Year**, *Article Number*, Page Range.

ISBN 978-3-03943-364-3 (Hbk)
ISBN 978-3-03943-365-0 (PDF)

Cover image courtesy of Davide Maggioni.

Contents

About the Editor . . . . **vii**

Preface to "Diversity of Coral-Associated Fauna" . . . . **ix**

Simone Montano
The Extraordinary Importance of Coral- Associated Fauna
Reprinted from: *Diversity* **2020**, *12*, 357, doi:10.3390/d12090357 . . . . **1**

Yee Wah Lau and James Davis Reimer
Zooxanthellate, Sclerite-Free, and Pseudopinnuled Octocoral *Hadaka nudidomus* gen. nov. et sp.
nov. (Anthozoa, Octocorallia) from Mesophotic Reefs of the Southern Ryukyus Islands
Reprinted from: *Diversity* **2019**, *11*, 176, doi:10.3390/d11100176 . . . . **5**

Tory J Chase and Mia O Hoogenboom
Differential Occupation of Available Coral Hosts by Coral-Dwelling Damselfish
(Pomacentridae) on Australia's Great Barrier Reef
Reprinted from: *Diversity* **2019**, *11*, 219, doi:10.3390/d11110219 . . . . **19**

**Davide Maggioni, Luca Saponari, Davide Seveso, Paolo Galli, Andrea Schiavo, Andrew N.
Ostrovsky and Simone Montano**
Green Fluorescence Patterns in Closely Related Symbiotic Species of *Zanclea*
(Hydrozoa, Capitata)
Reprinted from: *Diversity* **2020**, *12*, 78, doi:10.3390/d12020078 . . . . **39**

**Bert W. Hoeksema, Jaaziel E. García-Hernández, Godfried W.N.M. van Moorsel, Gabriël
Olthof and Harry A. ten Hove**
Extension of the Recorded Host Range of Caribbean Christmas Tree Worms (*Spirobranchus* spp.)
with Two Scleractinians, a Zoantharian, and an Ascidian
Reprinted from: *Diversity* **2020**, *12*, 115, doi:10.3390/d12030115 . . . . **57**

Bert W. Hoeksema and Jaaziel E. García-Hernández
Host-related Morphological Variation of Dwellings Inhabited by the Crab *Domecia acanthophora*
in the Corals *Acropora palmata* and *Millepora complanata* (Southern Caribbean)
Reprinted from: *Diversity* **2020**, *12*, 143, doi:10.3390/d12040143 . . . . **61**

**Javier Montenegro, Bert W. Hoeksema, Maria E. A. Santos, Hiroki Kise and James Davis
Reimer**
Zoantharia (Cnidaria: Hexacorallia) of the Dutch Caribbean and One New Species of
Parazoanthus
Reprinted from: *Diversity* **2020**, *12*, 190, doi:10.3390/d12050190 . . . . **65**

**Simone Montano, James D. Reimer, Viatcheslav N. Ivanenko, Jaaziel E. García-Hernández,
Godfried W.N.M. van Moorsel, Paolo Galli and Bert W. Hoeksema**
Widespread Occurrence of a Rarely Known Association between the Hydrocorals
Stylaster roseus and *Millepora alcicornis* at Bonaire, Southern Caribbean
Reprinted from: *Diversity* **2020**, *12*, 218, doi:10.3390/d12060218 . . . . **119**

**Ricardo González-Muñoz, Agustín Garese, Fabián H. Acuña, James D. Reimer and Nuno
Simões**
The Spotted Cleaner Shrimp, *Periclimenes yucatanicus* (Ives, 1891), on an Unusual
Scleractinian Host
Reprinted from: *Diversity* **2019**, *11*, 213, doi:10.3390/d11110213 . . . . **129**

About the Editor

Simone Montano is a researcher at the University of Milano-Bicocca, Department of Earth and Environmental Sciences (DISAT) in Italy, and Vice-Director of the Marine Research and High Education Center, Magoodhoo Island, Maldives. He is a marine biologist, mainly interested in the ecology and biology of the coral reef ecosystem. His current research activities focus on the assessment of coral health and diseases, with particular attention on new and emerging coral symbioses. All his activities are aimed to understand the dynamics that will drive this ecosystem under a climate change scenario, in order to develop and propose environmental management plans. To date, he has published more than 60 peer-reviewed papers in international scientific journals. He is a PADI diving instructor with an Advanced European Scientific Diver license, with >800 scientific dives and a total of >1000. His field work experiences include the Caribbean (St. Eustatius, Bonaire, and Curacao), the Indo-Pacific area (Maldives, Mauritius, Yemen, India, and Thailand), and the Red Sea (Egypt, Saudi Arabia).

Preface to "Diversity of Coral-Associated Fauna"

Mutualistic, commensalistic, and parasitic associations are extremely abundant in coral reef ecosystems. Reef-building corals are usually considered the most likely to provide numerous different habitats and to bear a huge number of symbiotic relationships. However, many other invertebrate groups such as sponges, bryozoans, and other cnidarians are known to establish strict symbiotic relationships with other marine organisms, even though their inter-specific interactions are poorly investigated. To date, symbiotic associations have mainly been studied by considering pairwise relationships, but in the vast majority of cases one host is typically inhabited by several other organisms (e.g., epibionts, commensals, and parasites) that may interact with each other and with the two partners. Unfortunately, even though these symbioses have been found to be more common than previously known, information regarding the nature, origin, and existence of any correlation with environmental factors is far from being fully elucidated.

In line with this, we believe that it is necessary to understand how these co-occurring organisms influence the symbiotic association considered, and how their combined effects influence the two partners. This information could be used to understand the mechanisms by which ecological interactions can mediate species' responses to disturbances and used to predict the ability of single organisms to persist in a rapidly changing environment.

For this reason, this book aims to explore the hidden diversity of coral reefs, focusing on some neglected components of the biodiversity of this extraordinary marine ecosystem, significantly improving our knowledge of the diversity, ecology, and role of the coral-associated fauna.

A special thanks goes to all authors of the papers published in this Special Issue. Their contributions regarding the ecological interactions in tropical coral reef ecosystems made by the combination of multidisciplinary approaches, taxonomic expertise, and dedicated biodiversity surveys revealed the existence of many previously unknown associations.

I strongly believe that similar in-depth studies addressed to identify and describe other hidden symbioses will be increasingly necessary in the future.

Simone Montano
Editor

Editorial

The Extraordinary Importance of Coral-Associated Fauna

Simone Montano [1,2]

[1] Department of Earth and Environmental Sciences (DISAT), University of Milan—Bicocca, Piazza della Scienza 1, 20126 Milan, Italy; simone.montano@unimib.it
[2] MaRHE Center (Marine Research and High Education Center), Magoodhoo Island, Faafu Atoll 12030, Maldives

Received: 14 September 2020; Accepted: 15 September 2020; Published: 16 September 2020

Abstract: Coral reefs are one of the most diverse marine ecosystems on Earth and one of the richest in terms of species interactions. Scleractinian corals are usually the most likely to provide numerous different habitats and to support many symbiotic relationships. However, many other invertebrate groups, such as sponges, bryozoans, and other cnidarians, establish strict symbiotic relationships with other marine organisms. Despite the nature of these relationships—as well as the factors that drive their establishment—being unclear in most cases, a few studies have already shown that some associations may increase the resistance of their hosts to external disturbances. Thus, the potential ability of each member of these diverse symbiotic assemblages to influence the fitness and long-term survival of their hosts bring the coral-associated fauna to the top of the list of coral reef studies. Unfortunately, the widespread degradation of coral reef ecosystems may threaten the existence of the intimate relationships that may go unrecognized complicating our understanding of the intricate networks connecting the fates of reef species. Therefore, this unprecedented loss of biodiversity calls for synergic conservation and monitoring actions aimed at significantly increasing our efforts to search for and describe as much of the diversity of coral-associated organisms as possible, shedding new light on the complex, elusive mechanisms controlling coral reef functioning.

Keywords: biodiversity; scleractinian; coral reefs; symbiosis; global change; impacts

---

Coral reefs encompass the highest biodiversity of any marine ecosystem of the planet [1]. This abundance is primarily due to the topographic complexity created by many benthic organisms, such as reef-building corals, sponges, bryozoans and other cnidarians that play a key role in creating the complex three-dimensional architecture of coral reef and providing a plethora of habitats to support an extraordinary diversity of organisms from all kingdoms of life [2].

The highly diverse fauna associated with these sessile reef organisms is dominated by invertebrates, belonging to numerous phyla—such as Arthropoda, Mollusca, Echinodermata, Anellida, Porifera and Cnidaria—depending on their hosts for food, refuges and habitats, and usually establishing strict symbiotic relationships in form of mutualistic, commensalistic and parasitic associations [3,4]. The coral-associated fauna assumes a considerable and unique importance considering that each member of these diverse symbiotic assemblages has the potential to influence the fitness and long-term survival of their host [2].

Reef-building corals, for example, are known to form associations with about a thousand of micro- and macro-organisms that, in many cases, appear to be strictly host specific. Despite the fact that the large number of them may contribute to the reduced health and mortality of corals through feeding or boring activities, many other species can be considered fundamental to the persistence and resilience of their host corals [4]. Indeed, more than 50% of coral-associated invertebrates are obligate coral dwellers, with some of them known to actively participate in nutrient recycling [5], to alleviate detrimental effects of sedimentation and actively defend colonies from coral-feeding organisms [6,7], or to slow down the progression of diseases as shown by the crabs of the genus *Cymo* [8]. More recently, coral symbiotic hydrozoans of the genus *Zanclea* has been proved to both reduce coral susceptibility to diseases and protect their hosts from predation [9], highlighting how far we are from the understanding of the mechanisms by which ecological interactions can mediate species' responses to disturbances.

Unfortunately, how many species are living on the coral reefs as well as the species of micro- and macroinvertebrates living in association with other reef organisms is still not clear. Most of the unknown reef communities consist of cryptofauna [10] that may be difficult to recognize in the field due to their tiny size [11,12], camouflage behavior [13,14], and because they live in habitats that are often overlooked, such us as caves, sediment or coral rubble [15], or because they are located in deep environments as the mesophotic zones [16]. This gap in knowledge can be exacerbated both in shallow and deep coral reefs if the parasites diversity is included since most species in most major parasite groups are still undiscovered or unnamed [17].

Bearing in mind the likely high degree of specialization and co-dependence of these symbiotic relationships, this lack of information appears dramatic in the light of the increasing number of threats contributing to the global decline of coral reefs [18]. Indeed, habitat degradation could have serious negative effects on the diversity of reefs and may disrupt these symbiotic relationships [19], intensifying the loss of biodiversity [20]. Thus, if preserving biodiversity is now considered a priority for any natural ecosystem, it is increasingly vital for the future of coral reefs in which thousands of coral-associated organisms could be negatively impacted by global change, on scales ranging from local declines to global extinction; these losses could have major downstream consequences for coral reef ecosystem function and stability [17].

The fundamental value of the papers published in this Special Issue is twofold. On one hand, it highlights the still-scarce knowledge of the ecological interactions in tropical coral reef ecosystems and the possible existence of many other so-far-unknown similar associations that deserve our attention. On the other hand, it highlights how the combination of multidisciplinary approaches, taxonomic expertise and dedicated biodiversity surveys can significantly improve our knowledge about the diversity, ecology and role of coral-associated fauna. Therefore, we hope that these studies can stimulate the exploration of neglected areas in reef ecology, increase significantly our effort in searching and describing as much the diversity of coral-associated organisms and systematically investigate the coral-associated biodiversity by adding coral-associated fauna surveys to largescale biodiversity monitoring programs.

Funding: This research received no external funding.

Conflicts of Interest: The authors declare no conflict of interest.

References

1. Fisher, R.; O'Leary, R.A.; Low-Choy, S.; Mengersen, K.; Knowlton, N.; Brainard, R.E.; Caley, M.J. Species Richness on Coral Reefs and the Pursuit of Convergent Global Estimates. *Curr. Biol.* **2015**, *25*, 500–505. [PubMed]
2. Gates, R.D.; Ainsworth, T.D. The nature and taxonomic composition of coral symbiomes as drivers of performance limit in scleractinian corals. *J. Exp. Mar. Biol. Ecol.* **2011**, *408*, 94–101.
3. Stella, J.S.; Jones, G.P.; Pratchett, M.S. Variation in the structure of epifaunal invertebrate assemblages among coral hosts. *Coral Reefs* **2010**, *29*, 957–973.
4. Stella, J.S.; Pratchett, M.S.; Hutchings, P.A.; Jones, G.P. Coral-associated invertebrates: Diversity, ecological importance and vulnerability to disturbance. *Oceanogr. Mar. Biol. Annu. Rev.* **2011**, *49*, 43–116.
5. Spotte, S. Supply of regenerated nitrogen to sea anemones by their symbiotic shrimp. *J. Exp. Mar. Biol.* **1996**, *198*, 27–36.
6. Stewart, H.L.; Holbrook, S.J.; Schmitt, R.J.; Brooks, A.J. Symbiotic crabs maintain coral health by clearing sediments. *Coral Reefs* **2006**, *25*, 609–615.

7. Rouzé, H.; Lecellier, G.; Mills, S.C.; Planes, S.; Berteaux-Lecellier, V.; Stewart, H. Juvenile *Trapezia* spp. crabs can increase juvenile host coral survival by protection from predation. *Mar. Ecol. Prog. Ser.* **2014**, *515*, 151–159.

8. Pollock, F.J.; Katz, S.M.; Bourne, D.G.; Willis, B.L. *Cymo melanodactylus* crabs slow progression of white syndrome lesions on corals. *Coral Reefs* **2013**, *32*, 43–48.

9. Montano, S.; Fattorini, S.; Parravicini, V.; Berumen, M.L.; Galli, P.; Maggioni, D.; Arrigoni, R.; Seveso, D.; Strona, G. Corals hosting symbiotic hydrozoans are less susceptible to predation and disease. *Proc. R. Soc. B* **2017**, *284*, 20172405. [CrossRef] [PubMed]

10. Reaka-Kudla, M.L. The global biodiversity of coral reefs: A comparison with rain forests. In *Biodiversity II: Understanding and Protecting our Natural Resources*; Reaka-Kudla, M.L., Ed.; Joseph Henry/National Academy Press: Washington, DC, USA, 1997; pp. 83–108.

11. Montano, S.; Arrigoni, R.; Pica, D.; Maggioni, D.; Puce, S. New insights into the symbiosis between *Zanclea* (Cnidaria, hydrozoa) and scleractinians. *Zool. Scripta* **2015**, *44*, 92–105. [CrossRef]

12. Ivanenko, V.N.; Hoeksema, B.W.; Mudrova, S.V.; Nikitin, M.A.; Martínez, A.; Rimskaya-Korsakova, N.N.; Berumen, M.L.; Fontaneto, D. Lack of host specificity of copepod crustaceans associated with mushroom corals in the Red Sea. *Mol. Phylogenetics Evol.* **2018**, *127*, 770–780.

13. Montano, S.; Maggioni, D. Camouflage of sea spiders (Arthropoda, Pycnogonida) inhabiting *Pavona varians*. *Coral Reefs* **2018**, *37*, 153.

14. Mehrotra, R.; Arnold, S.; Wang, A.; Chavanich, S.; Hoeksema, B.W.; Caballer, M. A new species of coral-feeding nudibranch (Mollusca: Gastropoda) from the Gulf of Thailand. *Mar. Biodiv.* **2020**, *50*, 36. [CrossRef]

15. Hoeksema, B.W. The hidden biodiversity of tropical coral reefs. *Biodiversity* **2017**, *18*, 8–12. [CrossRef]

16. Maggioni, D.; Montano, S.; Voigt, O.; Seveso, D.; Galli, P. A mesophotic hotel: The octocoral *Bebryce* cf. *grandicalyx* as a host. *Ecology* **2020**, *101*, e02950. [PubMed]

17. Carlson, C.J.; Hopkins, S.; Bell, K.C.; Doña, J.; Godfrey, S.S.; Kwak, M.L.; Lafferty, K.D.; Moir, M.L.; Speer, K.A.; Strona, G.; et al. A global parasite conservation plan. *Biol. Conserv.* **2020**, 108596.

18. Hughes, T.P.; Kerry, J.T.; Baird, A.H.; Connolly, S.R.; Chase, T.J.; Dietzel, A.; Hill, T.; Hoey, S.A.; Hoogenboom, M.O.; Jacobson, M.; et al. Global warming impairs stock–recruitment dynamics of corals. *Nature* **2019**, *568*, 387–390. [PubMed]

19. Caley, J.M.; Buckely, K.A.; Jones, G.P. Separating ecological effects of habitat fragmentation, degradation and loss of coral commensals. *Ecology* **2001**, *82*, 3435–3448. [CrossRef]

20. Kiers, E.; Palmer, T.; Ives, A.; Bruno, J.; Bronstein, J. Mutualisms in a changing world: An evolutionary perspective. *Ecol. Lett.* **2010**, *13*, 1459–1474. [CrossRef] [PubMed]

 *diversity*

Article

Zooxanthellate, Sclerite-Free, and Pseudopinnuled Octocoral *Hadaka nudidomus* gen. nov. et sp. nov. (Anthozoa, Octocorallia) from Mesophotic Reefs of the Southern Ryukyus Islands

Yee Wah Lau [1],* and James Davis Reimer [1,2]

[1] Molecular Invertebrate Systematics and Ecology Laboratory, Graduate School of Engineering and Science, University of the Ryukyus, 1 Senbaru, Nishihara, Okinawa 903-0213, Japan; jreimer@sci.u-ryukyu.ac.jp

[2] Tropical Biosphere Research Center, University of the Ryukyus, 1 Senbaru, Nishihara, Okinawa 903-0213, Japan

* Correspondence: lauyeewah87@gmail.com; Tel.: +31-6-3966-2659

http://zoobank.org/urn:lsid:zoobank.org:pub:1AB2F0C1-FAB0-40B0-AB7A-C07A296E9C50

Received: 25 August 2019; Accepted: 18 September 2019; Published: 22 September 2019

Abstract: Shallow water coral reefs are the most diverse marine ecosystems, but there is an immense gap in knowledge when it comes to understanding the diversity of the vast majority of marine biota in these ecosystems. This is especially true when it comes to understudied small and cryptic coral reef taxa in understudied ecosystems, such as mesophotic coral reef ecosystems (MCEs). MCEs were reported in Japan almost fifty years ago, although only in recent years has there been an increase in research concerning the diversity of these reefs. In this study we describe the first stoloniferous octocoral from MCEs, *Hadaka nudidomus* gen. nov. et sp. nov., from Iriomote and Okinawa Islands in the southern Ryukyus Islands. The species is zooxanthellate; both specimens host *Cladocopium* LaJeunesse & H.J.Jeong, 2018 (formerly *Symbiodinium* 'Clade C') and were collected from depths of ~33 to 40 m. Additionally, *H. nudidomus* gen. nov. et sp. nov. is both sclerite-free and lacks free pinnules, and both of these characteristics are typically diagnostic for octocorals. The discovery and morphology of *H. nudidomus* gen. nov. et sp. nov. indicate that we still know very little about stoloniferous octocoral diversity in MCEs, their genetic relationships with shallower reef species, and octocoral–symbiont associations. Continued research on these subjects will improve our understanding of octocoral diversity in both shallow and deeper reefs.

Keywords: *Cladocopium*; cryptofauna; marine biodiversity; mesophotic coral reef environments (MCEs); Octocorallia; stoloniferous octocorals; Symbiodiniaceae; taxonomy

---

1. Introduction

Coral reefs make up only 0.2% of the earth's ocean but are estimated to harbor a quarter of all marine species [1,2] and are the most diverse marine ecosystems on the planet. Unfortunately, these diverse marine communities are also one of the most threatened [3–6]. The 'hotspot' concept, a term used to mark a relatively restricted geographic area accommodating exceptionally high concentrations of biodiversity and endemism [7–9] has highlighted the wealth of species that are at risk and how localized such areas of richness can be [10]. However, there are vast gaps in knowledge concerning the majority of marine biota [11,12], making the recognition of biodiversity geographic patterns and hotspots questionable [13,14], as priorities identified for one taxon may not reflect the diversity of other taxa [14,15]. This is especially true for understudied localities and environments, such as understudied coral reef ecosystems.

Mesophotic coral reef ecosystems (MCEs) occur at depths below 30–40 m to 100 m or deeper in tropical and sub-tropical regions [16–19]. MCEs are considered understudied, as their depths make them difficult to access via normal SCUBA technology, yet too shallow for most submersibles [19,20]. However, research regarding MCEs has increased in recent years, along with calls for increased awareness and protection of these ecosystems [21]. Additionally, studies have demonstrated that MCEs can accommodate high levels of endemism [19,22] and harbor distinct geographical communities [19].

The coral reefs of southern Japan are at the top of the list in terms of global marine conservation priority, when considering the region's high levels of multi-taxon endemism and the high risk of biodiversity loss due to overexploitation and coastal development [23]. The Ryukyus Islands (RYS), i.e., Ryukyu Archipelago, encompass the southernmost region of Japan and include islands of different geological formations, ages, and sizes [24,25]. The surrounding waters and coral reefs fringing the islands are strongly influenced by the warm water brought from tropical areas around the Philippine islands by the Kuroshio Current, which flows towards the north along the west side of the island chain [24–26], extending warm water conditions northerly. As such, the RYS experience higher sea temperatures compared to other areas at similar latitudes, such as eastern Australia [27,28], thus creating unique coral reef conditions. Serious taxonomic and geographic biases are present in marine biodiversity research in the RYS. Most work in the RYS has been conducted on the phyla Pisces, Crustacea, and Cnidaria, with the majority of research on hermatypic hard corals (Scleractinia) and, surprisingly, far less work on other commercially important groups such as Echinodermata and Mollusca, as well as on other understudied small and cryptic coral reef taxa [25].

One such understudied small and cryptic group are octocorals belonging to the subordinal group, Stolonifera. Stoloniferan octocorals are characterized by having relatively simple colony growth forms, where the polyps are united basally by ribbon-like stolons, instead of being embedded side by side within a common coenenchymal mass [29–31]. There are seven families that are considered to belong to Stolonifera: Acrossotidae Bourne, 1914; Arulidae McFadden & Van Ofwegen, 2012; Clavulariidae Hickson, 1894; Coelogorgiidae Bourne, 1900; Cornulariidae Dana, 1846; Pseudogorgiidae Utinomi & Harada, 1973; and Tubiporidae Ehrenberg, 1828. The most speciose as well as the most studied family is Clavulariidae, which comprises approximately 30 genera and over 60 species. Until recently, all other families are all either monospecific or monogeneric, with no more than a few described species; recent studies have additionally introduced new genera and species for Arulidae [32,33], which is the most recently erected family.

Stoloniferous octocorals often have inconspicuous small colonies and polyps, which makes them hard to detect [32–34]. There are critical gaps that remain in the understanding of the functional and ecological significance of octocoral–zooxanthellae symbioses [35]. To date, only a handful of data are available on stoloniferous octocoral–symbiont relationships, which all concern members of the speciose genus *Clavularia* Blainville, 1830. *Clavularia* spp. from Australia all hosted *Durusdinium* LaJeunesse, 2018 [36,37]. On one other occasion, a single *Clavularia* sp. specimen from the Caribbean was found to host *Durusdinium* [38].

Obligate mutualistic symbioses play important roles in extending available energy resources and thus potentially influence biodiversity on reefs [36,39]; however, stoloniferous octocorals and their host–symbiont associations are a relatively underexamined fauna in the RYS, particularly from within MCEs. In this study we formally describe the zooxanthellate, sclerite-free, and pseudopinnuled octocoral *Hadaka nudidomus* gen. nov. et sp. nov. from MCEs around Okinawa and Iriomote Islands.

2. Materials and Methods

2.1. Specimen Collection and Morphological Examinations

One specimen was collected from one location each around Okinawa (August 2017; 26.856412 N, 128.245093 E) and Iriomote (December 2016; 24.370413 N, 123.736428 E) Islands (Figure 1). The specimens were found at depths of 33 and 40 m, respectively, by means of SCUBA (atmospheric air) and were preserved in 70–90% ethanol and subsamples in 95% ethanol. The current study is part of an ongoing survey of mesophotic and deep reef work. Vouchers and type material were deposited at the National Museum of Nature and Science (NSMT), Tokyo, Japan (Table 1). Both specimens were examined for the presence of sclerites by dissolving entire polyps and stolons in 4% hypochlorite (household bleach). Additionally, to visualize polyp tentacles and pseudopinnules, polyps were fixed in 20% formalin and embedded in methylene blue (1%).

2.2. DNA Extraction, Amplification, and Sequencing

DNA was extracted from polyps using a DNeasy Blood and Tissue kit (Qiagen, Tokyo, Japan). PCR amplification and sequencing were performed for four markers, of which three were mitochondrial (cytochrome c oxidase subunit I (COI), the MSH homologue mtMutS, and subunit ND6) and the fourth was the nuclear ribosomal marker (28S rDNA). Additionally, for Symbiodiniaceae, the nuclear internal transcribed spacer (ITS) region of ribosomal DNA was amplified. Protocols in [34] were followed and PCR products were treated with Exonuclease I and alkaline phosphate (shrimp) and sent for bidirectional sequencing on an ABI 3730XL (Fasmac, Kanagawa, Japan). Sequences were assembled and edited using Geneious R11 [40] and BioEdit [41]. COI, mtMutS, and ND6 were checked for introns, exons, and stop-codons in AliView [42].

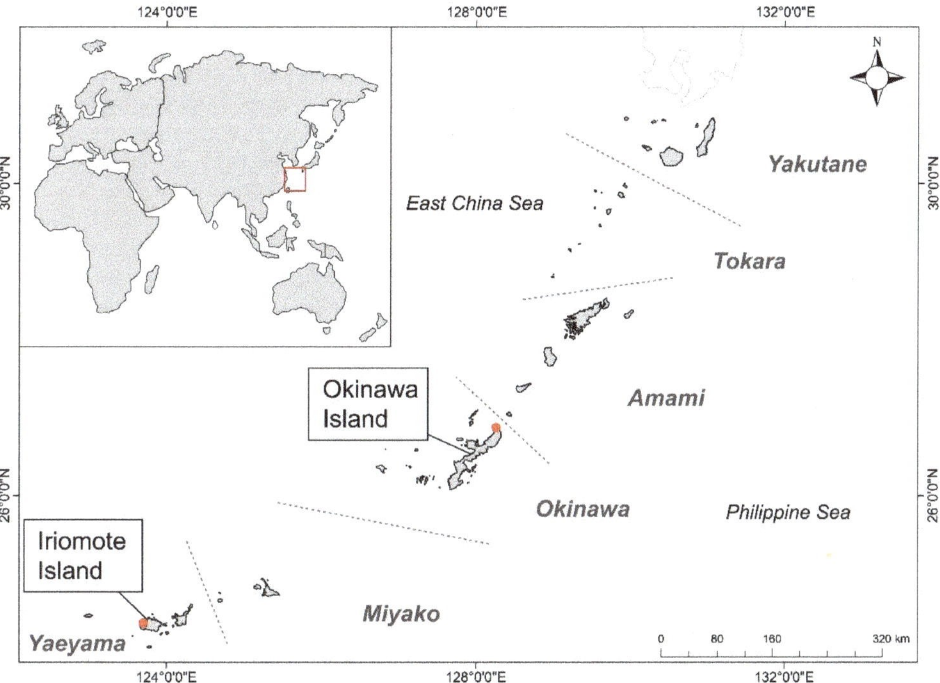

Figure 1. Map of the Ryukyus Islands (RYS), with the six island group divisions (grey dotted lines) and the two dive locations where *Hadaka nudidomus* gen. nov. et sp. nov. specimens were found (red dots) at Iriomote (NSMT-Co 1681, holotype) and Okinawa (NSMT-Co 1682, paratype) Islands.

Table 1. Overview of information on octocoral specimens collected from mesophotic coral reef ecosystems (MCEs) at Iriomote and Okinawa Islands, Okinawa Prefecture, Japan, including GenBank accession numbers and locality. Catalogue number: NSMT = National Museum of Nature and Science, Tokyo, Japan; n.a. = not available.

Family	Species	Catalogue Number	Locality/GPS (DMS)	Symbiodiniaceae Genus	GenBank Accession Numbers				
					28S rDNA	COI	mtMutS	ND6	ITS
Clavulariidae	*Hadaka nudidomus* gen. nov. et sp. nov.	NSMT-Co 1681 (holotype)	NE Uchibanare, Iriomote Isl./24.370413 N, 123.736428 E	*Cladocopium*	MN488601	MN488603	MN488605	n.a.	MN488607
	Hadaka nudidomus gen. nov. et sp. nov.	NSMT-Co 1682 (paratype)	Entrance Hedo Dome, Cape Hedo, Okinawa Isl./26.856412 N, 128.245093 E	*Cladocopium*	MN488602	MN488604	n.a.	MN488606	MN488608

2.3. Molecular Phylogenetic Analyses

Multiple sequence alignments were performed using MAFFT 7 [43] and coding markers were aligned using MACSE [44] under default parameters. The phylogenetic position of the collected specimens (n = 2) was determined by aligning the consensus sequences for markers 28S rDNA, COI, and mtMutS to a reference dataset of 124 octocoral genera, including *Cornularia pabloi* and *Cornularia cornucopiae* as outgroup (total n = 144), as used in Lau and Reimer [33]. This resulted in alignments of 887 bp for 28S rDNA, 717 bp for COI, and 714 bp for mtMutS, and a total concatenated three-marker dataset of 2318 bp. The separate markers were run in ML analyses, to check for contamination and congruency (Supplementary Materials Figures S1–S3).

A separate phylogenetic analysis was made to examine the lower level phylogenetic relationships of the collected mesophotic specimens, using a concatenated four-marker dataset. The concatenated four-marker dataset resulted in an alignment of 2670 bp (total n = 12). A total of seven reference species were included in the analysis, which clustered in nearby clades with the specimens in the three-marker dataset, including *Rhodelinda* sp. and *Telesto* sp. as outgroup. The four separate markers (28S rDNA, 787 bp; COI, 708 bp; mtMutS, 734 bp; ND6, 441 bp) were also run in ML analyses, to check for contamination and congruency (Supplementary Materials Figures S4–S7).

Additionally, ITS sequences from the two specimens were aligned with a total of 25 reference sequences (*Cladocopium* spp. and *Durusdinium* spp.), including *Gerakladium* sp. as outgroup. The resulting dataset comprised 641 bp and a total of 27 sequences and was run in ML analyses (Supplementary Materials Figure S8).

Alignments of the separate markers were concatenated using SequenceMatrix 1.8 [45]. ML analyses were run with RAX-ML 8 [46], using the GTRCAT model. The best ML tree was calculated using the –D parameter. A multi-parametric bootstrap search was performed, which automatically stopped based on the extended majority rule criterion. The Bayesian inference was performed with ExaBayes 1.5 [47] using the GTR substitution model. Four independent runs were run for 10,000,000 generations during which convergence (with a standard deviation of split frequencies < 2%) was reached. Bootstrap supports and posterior probabilities were depicted on the branches of the best ML tree using P4 [48]. The resulting trees were visualized in FigTree 1.4.2 [49]. Additionally, average distance estimations within species and within genera were computed using MEGA X [50] by analyzing pairwise measures of genetic distances (uncorrected P) among sequences (Supplementary Materials Tables S1–S3).

3. Systematic Account

Class Anthozoa
Subclass Octocorallia Ehrenberg, 1831
Order Alcyonacea Lamouroux, 1812
Family Clavulariidae Hickson, 1894

3.1. Genus *Hadaka* gen. nov.

Type species: *Hadaka nudidomus* sp. nov. by original designation and monotype.

Diagnosis: Colony with polyps connected through flattened ribbon-like stolons, which are loosely attached to a hard substrate. Polyps retract fully into the calyx, which is cylindrical to conical in shape, narrowing at the base and does not retract fully into the stolon. Tentacles have a wide rachis with a protruding ridge and pseudopinnules of different lengths arranged on either side, giving the polyps feather shaped tentacles. No sclerites. Zooxanthellate.

Remarks: *Hadaka* gen. nov. et sp. nov. shows gross resemblance to *Hanabira* Lau, Stokvis, Imahara & Reimer, 2019 in having a similar polyp shape with feather or petal shaped tentacles and fused pinnules, which can still be distinguished by shallow furrows. *Hadaka* gen. nov. et sp. nov. differs from *Hanabira* in having no sclerites in any part of the colony and having a protruding ridge on the upper side of the tentacle. Genetically, *Hadaka* gen. nov. is well-supported and positioned in a different phylogenetic clade from *Hanabira*. The closest sister taxa of *Hadaka* gen. nov. is *Acrossota* Bourne, 1914, which is also sclerite-free, but morphologically very different; *Acrossota* lacks pinnules completely.

Etymology: From the Japanese word *hadaka* (裸), meaning naked, bare, nude; denoting the absence of two characteristic features of octocorals, sclerites, and free pinnules. Gender: feminine.

http://zoobank.org/39430672-5ADA-4EFF-9F5A-B4076B6B90C0

3.2. *Hadaka nudidomus* sp. nov.

See Figure 2.

Material examined: All specimens were collected from Okinawa Prefecture, Japan. *Holotype*: NSMT-Co 1681, northeast Uchibanare, Iriomote Island (24.370413 N,123.736428 E), ~40 m depth, 19 December 2016, coll. D. Uyeno. GenBank accession numbers: 28S rDNA, MN488601; COI, MN488603; mtMutS, MN488605. *Paratype*: NSMT-Co 1682, entrance to Hedo Dome, Cape Hedo, Okinawa Island (26.856412 N, 128.245093 E), 33 m depth, 18 August 2017, coll. J.D. Reimer. GenBank accession numbers: 28S rDNA, MN488602; COI, MN488604; ND6, MN488606.

Description: Holotype colony consists of 15 polyps with flattened ribbon-like stolons encrusting a sponge. Polyps can be seen individually or clustered in groups and are spaced apart irregularly, 3 mm to 2 cm in between polyps and clusters. Stolons are 0.5 mm at their narrowest and 1 mm at their widest point. Polyps retract fully into the calyx (~1.8 mm wide and ~3.55 mm in length), which is cylindrical to conical shaped, narrowing at the base, and does not retract fully into the stolon. Expanded polyps are ~4–5 mm diameter in life. Tentacles have a wide rachis with a protruding ridge on the upper side and long pseudopinnules arranged on either side (~24–26 pseudo-pairs), giving the polyps feather shaped tentacles. When stained with methylene blue, the outline of the tentacles can be observed. Structures of the pinnule axis are visible; however, the notches that distinguish the pseudopinnules are not observed in the contour of the tentacle (Figure 2d). No sclerites were found in any parts of the specimens. Polyps are brown in life and yellowish-white in ethanol (Figure 2c). Zooxanthellate.

Morphological variation: There is a difference in color between the polyps of the holotype (NSMT-Co 1681) and paratype (NSMT-Co 1682); the polyps of the holotype are brown with a white oral disc and base of the tentacles and the polyps of the paratype are whitish yellow with a bright blue oral disc (Figure 2a,b).

Distribution: Southwestern Japan, southern Ryukyus Islands, around northern Okinawa Island, and inside the bay of western Iriomote Island in the East China Sea. Specimens were collected from depths of ~33–40 m.

Remarks: The polyps of paratype NSMT-Co 1682 were all used for DNA extraction and sclerite examination, as they were initially thought to be a *Hanabira yukibana* specimen; three fragments of rock with stolon remain. The holotype colony (NSMT-Co 1681) was attached to sponge tissue, but this epibiont is not obligate, as the paratype was attached to rock.

Habitat: The holotype (NSMT-Co 1681) was found attached to sponge on a large piece of coral rubble (>15 cm) lying on a mixed small rubble/soft sediment bottom. The paratype (NSMT-Co 1682) was found on consolidated hard carbonate bottom. Both colonies were on the upward-facing side of the bottom.

Etymology: From Latin *nudus*, meaning naked or bare, and *domus*, meaning home or house; denoting the 'naked' host habitat in which the zooxanthellae reside, as the species is sclerite-free.

http://zoobank.org/71620752-8C33-4DCE-9B6E-DD7FC2DA3E20

4. Molecular Results

This study added a total of six sequences of *Hadaka nudidomus* gen. nov. et sp. nov. to the public reference database GenBank and no barcodes were available before. For the family Symbiodiniaceae, two *Cladocopium* spp. sequences were added. The phylogenies resulting from the ML analyses of the separate markers (COI, 28S rDNA, mtMutS, ND6) were highly congruent with those from the concatenated alignments for both the three- and four-marker datasets (Supplementary Materials Figures S1–S7). ML and Bayesian analyses for the concatenated datasets yielded almost identical tree phylogenies (Supplementary Materials Figure S9). Sequences of *Hadaka nudidomus* gen. nov. et sp. nov. collected from Okinawa and Iriomote Islands formed a completely-supported clade, containing sclerite-free species only: species of clavulariid genus *Phenganax* Alderslade & McFadden, 2011 and monospecific acrossotid genus *Acrossota* Bourne, 1914 in both the three- and four-marker analyses (Figures 3 and 4).

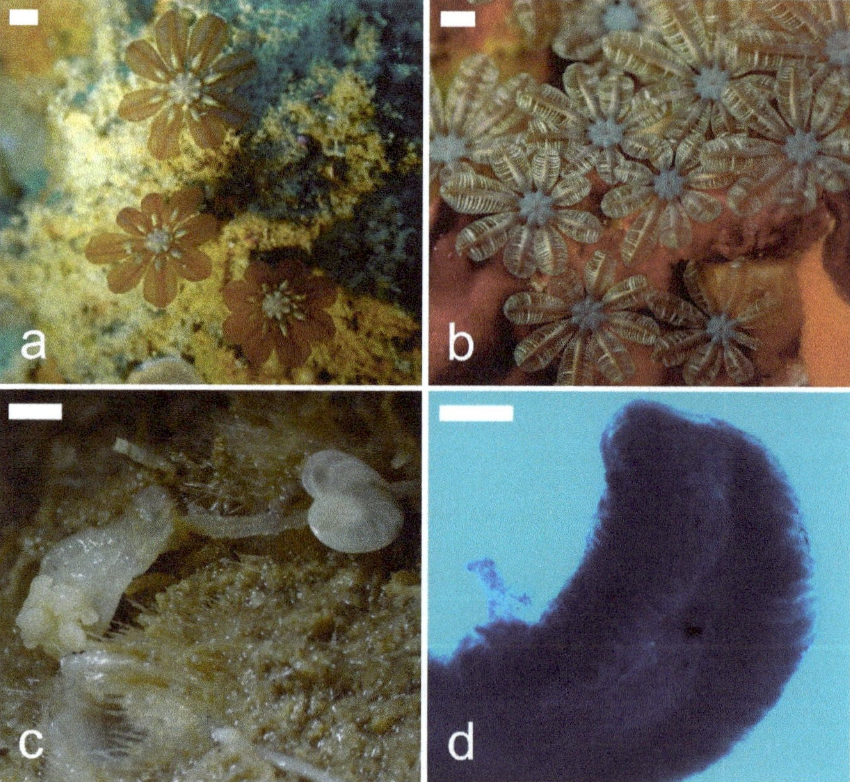

Figure 2. Photographs of *Hadaka nudidomus* gen. nov. et sp. nov.: (**a**) in situ holotype NSMT-Co 1681, scale bar approximately 1 mm; (**b**) in situ paratype NSMT-Co 1682, scale bar approximately 1 mm; (**c**) holotype in ethanol, scalebar 1 mm; (**d**) holotype in methylene blue staining, scale bar 0.1 mm.

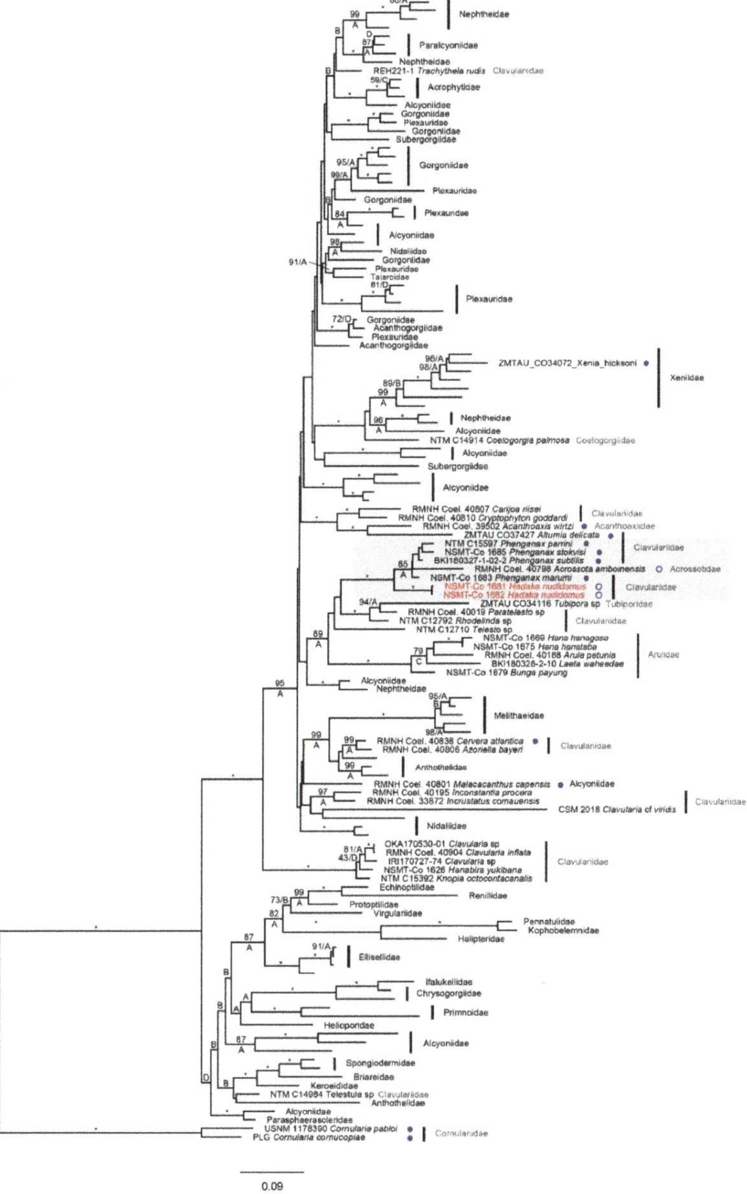

0.09

Figure 3. Phylogenetic relationships among 122 octocoral genera (total *n* = 144), including two species, *Hadaka nudidomus* gen. nov. et sp. nov. (highlighted red), collected at Iriomote and Okinawa Islands using the combined 28S rDNA + COI + mtMutS dataset. The best maximum likelihood tree is shown, with values at branches representing bootstrap probabilities (shown when >70%; top/left) and Bayesian posterior probabilities (shown when >0.80; bottom/right; A = 1.00, B = 0.95–0.99, C = 0.90–0.94, D = 0.80–0.89). * represents 100%/1.00 for both analyses. Non-stoloniferous families are shown with family classification only and stoloniferous families are highlighted in grey. Sclerite-free species are indicated with a blue dot. Species that are both sclerite-free and lack free pinnules are indicated with a blue circle. *Cornularia* spp. were used as outgroup.

Hadaka gen. nov. was sister to the remaining clade in the three-marker analyses; however, in the four-marker analyses, *Phenganax* was sister to *Hadaka* gen. nov. and *Acrossota*. Nonetheless, in both phylogenies, the two specimens of *Hadaka* gen. nov. formed a completely-supported clade.

Additionally, genetic distances gave further support to phylogenetic affinities and morphological features justifying the establishment of a new genus. Between-genus distances (*Hadaka* compared to *Acrossota* and *Phenganax*) for COI were 2.52–2.54% and 7.69–13.33% for mtMutS, which are well above the intergeneric range for octocorals [51]. Additional comparisons between *Hadaka nudidomus* sp. nov. specimens, *Acrossota amboinenesis* Burchardt, 1902 and *Phenganax* spp.; *Phenganax parrini* Alderslade & McFadden, 2011, *Phenganax marumi* Lau & Reimer, 2019, *Phenganax subtilis* Lau & Reimer, 2019, *Phenganax stokvisi* Lau & Reimer, 2019, also resulted in ranges (COI: 2.15–2.97%, MSH: 7.00–13.33%) that indicated that *Hadaka* gen. nov. specimens belong to a different genus (Supplementary Materials Tables S1–S3). There were no differences (0%) when comparing genetic distances within the two *Hadaka* specimens, indicating that the specimens are of the same species.

Hadaka nudidomus gen. nov. et sp. nov. specimens were analyzed for the presence of zooxanthellae, and identical sequences of Symbiodiniaceae were found. Both *Hadaka nudidomus* specimens collected from Okinawa and Iriomote Island hosted *Cladocopium* LaJeunesse & H.J.Jeong, 2018 (formerly *Symbiodinium* 'Clade C').

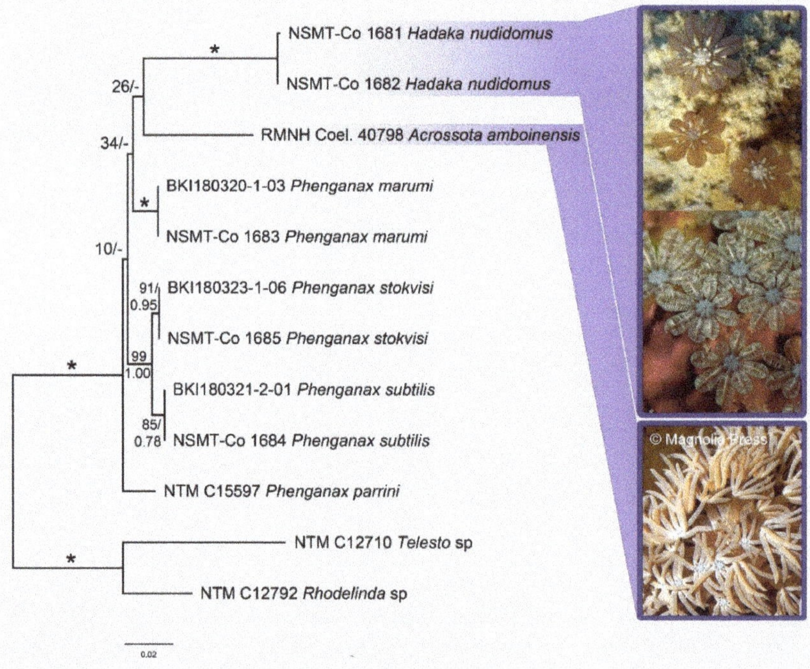

Figure 4. Phylogenetic reconstruction using a four-marker concatenated dataset (28S rDNA + COI + mtMutS + ND6) among *Hadaka nudidomus* gen. nov. et sp. nov., closest sister species *Phenganax* spp. and outgroup specimens *Rhodelinda* sp. and *Telesto* sp. (total *n* = 12). The best maximum likelihood tree is shown, with values at branches representing bootstrap probabilities in percentages (top/left) and Bayesian posterior probabilities (bottom/right). In situ photographs are shown for the two octocoral species that are sclerite-free and lack free pinnules, *Hadaka nudidomus* gen. nov. et sp. nov. and *Acrossota amboinensis*. Photograph credit: in situ image RMNH Coel. 40798 *Acrossota amboinensis*, by Daniel Knop (modified from [52]; reproduced with permission from copyright holder).

5. Discussion

In the three- and four-marker phylogenies, there was disparity in the position of *Hadaka nudidomus* gen. nov. et sp. nov. and *Acrossota amboinensis*. It remains unresolved how these genera and genus *Phenganax* are related to one another. A possible explanation could be that there is no sufficient signal in the sequences of both *Hadaka nudidomus* gen. nov. et sp. nov. and *Acrossota amboinenesis* due to the fact that the closest relatives for these genera are yet to be discovered.

Morphologically, there was a difference between the coloration of the polyps of the holotype and paratype found at Iriomote and Okinawa Islands, respectively. The differences in coloration suggested that perhaps the specimens hosted different members of Symbiodiniaceae. However, both specimens hosted genus *Cladocopium* and thus no biogeographical distinction in Symbiodiniaceae was observed. Members of *Cladocopium* spp. are known to be adapted to a wide range of temperatures and irradiances [53], which would be expected from MCEs, where irradiances are not only subject to seasonal variations but are already reduced.

Hadaka nudidomus gen. nov. et sp. nov is the first zooxanthellate stoloniferous octocoral described from mesophotic depths. Only one other zooxanthellate octocoral, an alcyoniid species, *Sinularia mesophotica* Benayahu, McFadden, Shoham & van Ofwegen, 2017, has been explicitly described from mesophotic depths [54]. However, it was not further specified which genus or species of Symbiodiniaceae was hosted by *S. mesophotica* and therefore, we cannot yet hypothesize differences of zooxanthellae hosted by octocorals from MCEs.

Nonetheless, another recent study has shown that there are geographical differences in the genera of Symbiodiniaceae in *Hanabira yukibana* Lau, Stokvis, Imahara & Reimer, 2019 from shallow coral reefs, as specimens found from Okinawa Island hosted *Cladocopium* while *Durusdinium* LaJeunesse, 2018 was hosted in specimens from Iriomote Island. However, in this previous study, similar to the current study, no consistently different patterns of polyp coloration related to symbiont associations were observed [34]. To this end, finer-scale examinations of Symbiodiniaceae using faster-evolving DNA markers [55] may reveal patterns yet unseen.

Hadaka nudidomus gen. nov. et sp. nov. is the second species within Octocorallia after *Acrossota amboinensis* that has no sclerites in any part of the colony and also no free pinnules; both species are taxonomically placed within family Clavulariidae. *Acrossota amboinensis* differs from *Hadaka nudidomus* gen. nov. et sp. nov. in colony form, polyp morphology, and habitat (Figure 4); *A. amboinensis* does not have pseudopinnules, but instead lacks pinnules completely and has, so far, not been found at mesophotic depths. When comparing *A. amboinensis* to *Phenganax* spp. there are also distinct morphological differences; all *Phenganax* species have free pinnules and have completely different polyp and tentacle shapes. Nonetheless, the genera *Acrossota* and *Phenganax* are phylogenetically closely related. In a recent study, the phylogenetic topology for these genera was different from that generated in the current study [33], in which *Acrossota* is placed basally to all *Phenganax* species. As a result of the unresolved phylogenetic location of *Acrossota* within Clavulariidae, while it is clear these three genera are distinct, it remains unclear how *Hadaka* and *Phenganax* are related to *Acrossota*.

Moreover, it can be concluded that several octocoral species lack both sclerites and free pinnules, and thus, that such features are not completely rare in octocorals, which raises important implications for the definition of subclass Octocorallia, as sclerite characterization and the presence of pinnated tentacles are two of the major diagnostic features of the group [52,56,57].

It is clear that more species diversity data from many marine regions are needed before we can state with certainty that the southern Ryukyus harbor high levels of stoloniferous octocoral diversity and endemism, but at least it can be said that this region potentially harbors many undiscovered species, not only in shallow coral reefs [33,34], but also among the many unexplored MCEs in this region.

Recent studies have shown that MCEs harbor distinct and independent biological communities when compared to shallower reefs [21]. MCEs are not only affected by anthropogenic and natural impacts as are shallow reefs but have seldom been the focus of specific conservation efforts [21,58].

Thus, researchers have only begun to scratch the surface of what we know about mesophotic marine life [21,58,59], including information on stoloniferous octocoral diversity and octocoral–zooxanthellae relationships in MCEs. The discovery of *Hadaka nudidomus* gen. nov. et sp. nov. and other recent discoveries [54,57] emphasize the need for continued studies on MCE octocoral diversity, as undescribed species may disappear before we have the opportunity to discover and study them [21].

Supplementary Materials: The following are available online at http://www.mdpi.com/1424-2818/11/10/176/s1, Figure S1: Maximum Likelihood phylogeny reconstruction of 28S rDNA gene region of *Hadaka nudidomus* gen. nov. et sp. nov. from Okinawa and Iriomote Islands (Japan) and octocoral references from 123 genera, including outgroup *Cornularia* spp., Figure S2: Maximum Likelihood phylogeny reconstruction of COI gene region of *Hadaka nudidomus* gen. nov. et sp. nov. from Okinawa and Iriomote Islands (Japan) and octocoral references from 123 genera, including outgroup *Cornularia* spp., Figure S3: Maximum Likelihood phylogeny reconstruction of mtMutS gene region of *Hadaka nudidomus* gen. nov. et sp. nov. from Okinawa and Iriomote Islands (Japan) and octocoral references from 123 genera, including outgroup *Cornularia* spp., Fiugre S4: Maximum Likelihood phylogeny reconstruction of 28S rDNA gene region of *Hadaka nudidomus* gen. nov. et sp. nov. from Okinawa and Iriomote Islands (Japan) and five octocoral references (*Phenganax* spp., *Acrossota amboinensis*), and outgroup (*Telesto* sp., *Rhodelinda* sp.)., Figure S5: Maximum Likelihood phylogeny reconstruction of COI gene region of *Hadaka nudidomus* gen. nov. et sp. nov. from Okinawa and Iriomote Islands (Japan) and five octocoral references (*Phenganax* spp., *Acrossota amboinensis*), and outgroup (*Rhodelinda* sp.)., Figure S6: Maximum Likelihood phylogeny reconstruction of mtMutS gene region of *Hadaka nudidomus* gen. nov. et sp. nov. from Okinawa and Iriomote Islands (Japan) and four octocoral references (*Phenganax* spp., *Acrossota amboinensis*), and outgroup (*Telesto* sp.)., Figure S7: Maximum Likelihood phylogeny reconstruction of ND6 gene region of *Hadaka nudidomus* gen. nov. et sp. nov. from Okinawa and Iriomote Islands (Japan) and two octocoral references (*Phenganax* spp.)., Figure S8: Maximum likelihood phylogenetic reconstruction of gene region ITS of Symbiodiniaceae hosted by *Hadaka nudidomus* gen. nov. et sp. nov. specimens from Okinawa and Iriomote Islands (Japan) and reference taxa *Durusdinium* sp. (= former *Symbiodinium* 'Clade D', n = 14) and *Cladocopium* sp. (former Symbiodinium 'Clade C', n = 10) and outgroup sister taxa, *Gerakladium* sp. (= former *Symbiodinium* 'Clade G') as used in Lau et al (2019)., Figure S9: Bayesian inference phylogeny reconstruction of the combined 28S rDNA+COI+mtMutS gene regions of *Hadaka nudidomus* gen. nov. et sp. nov. from Okinawa and Iriomote Islands (Japan) and octocoral references from 123 genera, including outgroup *Cornularia* spp., Table S1: Number of base differences per site from averaging over all sequence pairs between stoloniferous octocoral genera (*Hadaka* gen. nov., *Phenganax*, *Acrossota*) is shown (p expressed as percentage) for COI and mtMutS gene regions. Standard error estimates (S.E.) are shown above the diagonal. Analysis involved 9 and 6 nucleotide sequences for COI and mtMutS, respectively. All positions containing gaps and missing data were eliminated. There were totals of 708 and 734 positions in the final dataset for COI and mtMutS, respectively. Evolutionary analyses were conducted in MEGA X (Kumar et al. 2018)., Table S2: Number of base differences per site from averaging over all sequence pairs between stoloniferous octocoral taxa (*Hadaka nudidomus* gen. nov. sp. nov., *Phenganax* spp., *Acrossota amboinensis*) is shown (p expressed as percentage) for gene regions COI and mtMutS. Standard error estimates (S.E.) are shown above the diagonal. Analysis involved 7 nucleotide sequences for both COI and mtMutS. All positions containing gaps and missing data were eliminated. There were totals of 717 and 881 positions in the final dataset for COI and mtMutS, respectively. Evolutionary analyses were conducted in MEGA X (Kumar et al. 2018)., Table S3: Estimates of average evolutionary divergence over sequence pairs within stoloniferous octocoral genera (*Hadaka* gen. nov., *Phenganax*, *Acrossota*, *Rhodelinda*, *Telesto*) for gene regions COI and mtMutS. The numbers of base differences per site from averaging over all sequence pairs within each group (d) are shown (p expressed as percentage). Standard error estimates (S.E.) are shown in the second column and were obtained by a bootstrap procedure (1000 replicates). Analyses involved 9 and 6 nucleotide sequences for COI and mtMutS, respectively. All positions containing gaps and missing data were eliminated. There were totals of 708 and 734 positions in the final dataset for COI and mtMutS, respectively. Evolutionary analyses were conducted in MEGA X (Kumar et al. 2018).

Author Contributions: Conceptualization, methodology, and writing—review and editing, Y.W.L. and J.D.R.; formal analysis, investigation, data curation, writing—original draft preparation, and visualization, Y.W.L.; supervision and resources, J.D.R.

Funding: This research received no external funding.

Acknowledgments: Daisuke Uyeno (Kagoshima University) is thanked for providing the holotype specimen NSMT-Co 1681. Tohru Naruse (University of the Ryukyus) is thanked for logistical support in the field in Iriomote Island. Kristen G.Y. Soong (University of the Ryukyus) is thanked for her help with laboratory work. Frank R. Stokvis (Naturalis Biodiversity Center) is thanked for providing help with phylogenetic analyses.

Conflicts of Interest: The authors declare no conflict of interest.

References

1. Buddemeier, R.W.; Kleypas, J.A.; Aronson, R.B. *Coral Reefs and Global Climate Change: Potential Contributions of Climate Change to Stresses on Coral Reef Ecosystems, Report*; Pew Centre on Global Climate Change: Arlington, VA, USA, 2004; 56p, Available online: http://www.reefresilience.org/pdf/Coral_Reefs_and_Global_Climate_Change.pdf (accessed on 2 April 2019).
2. Chen, P.Y.; Chen, C.C.; Chu, L.; McCarl, B. Evaluating the economic damage of climate change. *Glob. Environ. Chang.* **2015**, *30*, 12–20. [CrossRef]
3. Hughes, T.P.; Baird, A.H.; Bellwood, D.R.; Card, M.; Connolly, S.R.; Folke, C.; Grosberg, R.; Hoegh-Guldberg, O.; Jackson, J.B.C.; Kleypas, J.; et al. Climate change, human impacts, and the resilience of coral reefs. *Science* **2003**, *301*, 929–933. [CrossRef] [PubMed]
4. Pandolfi, J.M.; Bradbury, R.H.; Sala, E.; Hughes, T.P.; Bjorndal, K.A.; Cooke, R.G.; McArdle, D.; McClenachan, L.; Newman, M.J.H.; Paredes, G.; et al. Global trajectories of the long-term decline of coral reef ecosystems. *Science* **2003**, *301*, 955–958. [CrossRef] [PubMed]
5. Meyer, C.P.; Geller, J.B.; Paulay, G. Fine scale endemism on coral reefs: Archipelagic differentiation in turbinid gastropods. *Evolution* **2005**, *59*, 113–125. [CrossRef] [PubMed]
6. Hughes, T.P.; Kerry, J.T.; Baird, A.H.; Connolly, S.R.; Dietzel, A.; Eakin, C.A.; Heron, S.F.; Hoey, A.S.; Hoogenboom, M.O.; Liu, G.; et al. Global warming transforms coral reef assemblages. *Nature* **2018**, *556*, 492–496. [CrossRef] [PubMed]
7. Mittermeier, R.A.; Myers, N.; Thomsen, J.B.; da Fonseca, G.A.B.; Olivieri, S. Biodiversity hotspots and major tropical wilderness areas: Approaches to setting conservation priorities. *Conserv. Biol.* **1998**, *12*, 516–520. [CrossRef]
8. Mittermeier, R.A.; Myers, N.; Mittermeier, C.G. *Hotspots—Earth's Biologically Richest and Most Endangered Terrestrial Ecoregions*; Cemex: Mexico City, Mexico, 1999; 431p.
9. Allen, G.R. Conservation hotspots of biodiversity and endemism for Indo-Pacific coral reef fishes. *Aquat. Conserv.* **2008**, *18*, 541–556. [CrossRef]
10. Bellwood, D.R.; Meyer, C.P. Searching for heat in a marine biodiversity hotspot. *J. Biogeogr.* **2009**, *36*, 569–576. [CrossRef]
11. Appeltans, W.; Ahyong, S.T.; Anderson, G.; Angel, M.V.; Artois, T.; Bailly, N.; Bamber, R.; Barber, A.; Bartsch, I.; Berta, A.; et al. The magnitude of global marine species diversity. *Curr. Biol.* **2012**, *22*, 2189–2202. [CrossRef]
12. Troudet, J.; Grandcolas, P.; Blin, A.; Vignes-Lebbe, R.; Legendre, F. Taxonomic bias in biodiversity data and societal preferences. *Sci. Rep.* **2017**, *7*, 9132. [CrossRef]
13. Mace, G.M.; Balmford, A.; Boitani, L.; Cowlishaw, G.; Dobson, A.P.; Faith, D.P. It's time to work together and stop duplicating conservation efforts. *Nature* **2000**, *405*, 393. [CrossRef] [PubMed]
14. Hooper, J.N.A.; Kennedy, J.A.; Quinn, R.J. Biodiversity 'hotspots', patterns of richness and endemism, and taxonomic affinities of tropical Australian sponges (Porifera). *Biodivers. Conserv.* **2002**, *11*, 851–885. [CrossRef]
15. Kerr, J.T. Species richness, endemism, and choice of areas for conservation. *Conserv. Biol.* **1997**, *11*, 1094–1100. [CrossRef]
16. Puglise, K.A.; Hinderstein, L.M.; Marr, J.C.A.; Dowgiallo, M.J.; Martinez, F.A. *Mesophotic Coral Ecosystems Research Strategy: International Workshop to Prioritize Research and Management Needs for Mesophotic Coral Ecosystems, Jupiter, Florida, 12–15 July 2008*; NOAA Technical Memorandum NOS NCCOS 98 and OAR OER 2; NOAA National Centers for Coastal Research, NOAA Undersea Research Program: Silver Spring, MD, USA, 2009; 24p.
17. Kahng, S.E.; Garcia-Sais, J.R.; Spalding, H.L.; Brokovich, E.; Wagner, D.; Weil, E.; Hinderstein, L.; Toonen, R.J. Community ecology of mesophotic coral reef ecosystems. *Coral Reefs* **2010**, *29*, 255–275. [CrossRef]
18. Sinniger, F.; Harii, S. Studies on Mesophotic Coral Ecosystems in Japan. In *Coral Reef Studies of Japan. Coral Reefs of the World*; Iguchi, A., Hongo, C., Eds.; Springer: Singapore, 2018; Volume 13, pp. 149–162.
19. Laverick, J.H.; Piango, S.; Andradi-Brown, D.A.; Exton, D.A.; Bongaerts, P.; Bridge, T.C.; Lesser, M.P.; Pyle, R.L.; Slattery, M.; Wagner, D.; et al. To what extent do mesophotic coral ecosystems and shallow reefs share species of conservation interest? A systematic review. *Environ. Evid.* **2018**, *7*, 15. [CrossRef]

20. Pyle, R.L. *Ocean Pulse: A Critical Diagnosis*; Tanacredi, J.T., Loret, J., Eds.; Springer: Boston, MA, USA, 1998; pp. 71–88.

21. Rocha, L.A.; Pinheiro, H.T.; Shepherd, B.; Papastamatiou, Y.P.; Luiz, O.J.; Pyle, R.L.; Bongaerts, P. Mesophotic coral ecosystems are threatened and ecologically distinct from shallow water reefs. *Science* **2018**, *361*, 281–284. [CrossRef] [PubMed]

22. Kosaki, R.K.; Pyle, R.L.; Leonard, J.C.; Hauk, B.B.; Whitton, R.K.; Wagner, D. 100% endemism in mesophotic reef fish assemblages at Kure Atoll, Hawaiian Islands. *Mar. Biodivers.* **2017**, *47*, 783–784. [CrossRef]

23. Roberts, C.M.; McClean, C.J.; Veron, J.E.N.; Hawkins, J.P.; Allen, J.R.; McAllister, D.E.; Mittermeier, C.G.; Schueler, F.W.; Spalding, M.; Wells, F.; et al. Marine biodiversity hotspots and conservation priorities for tropical reefs. *Science* **2002**, *295*, 1280–1285. [CrossRef]

24. Kizaki, K. Geology and tectonics of the Ryukyu Islands. *Tectonophysics* **1986**, *125*, 193–207. [CrossRef]

25. Reimer, J.D.; Biondi, P.; Lau, Y.W.; Masucci, G.; Nguyen, X.H.; Santos, M.E.A.; Wee, H.B. Marine biodiversity research in the Ryukyu Islands, Japan: Current status and trends. *PeerJ* **2019**, *7*, e6532. [CrossRef]

26. Andres, M.; Park, J.H.; Wimbush, M.; Zhu, X.H.; Chang, K.I.; Ichikawa, H. Study of the Kuroshio/Ryukyu current system based on satellite-altimeter and in situ measurements. *J. Oceanogr.* **2008**, *64*, 937–950. [CrossRef]

27. Veron, J.E.N. Conservation of biodiversity: A critical time for the hermatypic corals of Japan. *Coral Reefs* **1992**, *11*, 13–21. [CrossRef]

28. Rodriguez-Lanetty, M.; Hoegh-Guldberg, O. Symbiont diversity within the widespread scleractinian coral *Plesiastrea versipora*, across the northwestern Pacific. *Mar. Biol.* **2003**, *143*, 501–509. [CrossRef]

29. Fabricius, K.; Alderslade, P. *Soft Corals and Sea Fans: A Comprehensive Guide to the Tropical Shallow-Water Genera of the Central-West Pacific, the Indian Ocean and the Red Sea*; Australian Institute of Marine Science: Townsville, Australia, 2001.

30. Daly, M.; Brugler, M.R.; Cartwright, P.; Collins, A.G.; Dawson, M.N.; Fautin, D.G.; France, S.C.; McFadden, C.S.; Opresko, D.M.; Rodriguez, E.; et al. The phylum Cnidaria: A review of phylogenetic patterns and diversity 300 years after Linnaeus. *Zootaxa* **2007**, *1668*, 127–182.

31. McFadden, C.S.; van Ofwegen, L.P. Stoloniferous octocorals (Anthozoa, Octocorallia) from South Africa, with descriptions of a new family of Alcyonacea, a new genus of Clavulariidae, and a new species of *Cornularia* (Cornulariidae). *Invertebr. Syst.* **2012**, *26*, 331–356. [CrossRef]

32. Lau, Y.W.; Stokvis, F.R.; Van Ofwegen, L.P.; Reimer, J.D. Stolonifera from shallow waters in the north-western Pacific: A description of a new genus and two new species within the Arulidae (Anthozoa, Octocorallia). *ZooKeys* **2018**, *790*, 1–19. [CrossRef] [PubMed]

33. Lau, Y.W.; Reimer, J.D. A first phylogenetic study on stoloniferous octocorals off the coast of Kota Kinabalu, Sabah, Malaysia, and the description of three new genera and six new species. *ZooKeys* **2019**, *872*, 127–158. [CrossRef]

34. Lau, Y.W.; Stokvis, F.R.; Imahara, Y.; Reimer, J.D. The stoloniferous octocoral, *Hanabira yukibana*, gen. nov., sp. nov., of the southern Ryukyus has morphological and symbiont variation. *Contrib. Zool.* **2019**, *54–77*. [CrossRef]

35. Van de Water, J.A.J.M.; Allemand, D.; Ferrier-Pagès, C. Host-microbe interactions in octocoral holobionts-recent advances and perspectives. *Microbiome* **2018**, *6*, 64. [CrossRef]

36. Van Oppen, M.J.H.; Mieog, J.C.; Sánchez, C.A.; Fabricius, K.E. Diversity of alga symbionts (zooxanthellae) in octocorals: The roles of geography and host relationships. *Mol. Ecol.* **2005**, *14*, 2403–2417. [CrossRef]

37. Goulet, T.L.; Simmons, C.; Goulet, D. Worldwide biogeography of *Symbiodinium* in tropical octocorals. *Mar. Ecol. Prog. Ser.* **2008**, *355*, 45–58. [CrossRef]

38. LaJeunesse, T.C.; Bhagooli, R.; Hidaka, M.; de Vantier, L.; Done, T.; Schmidt, G.W.; Fitt, W.K.; Hoegh-Guldberg, O. Closely related *Symbiodinium* spp. differ in relative dominance in coral reef host communities across environmental, latitudinal and biogeographic gradients. *Mar. Ecol. Prog. Ser.* **2004**, *284*, 147–161. [CrossRef]

39. Fabricius, K.E.; De'ath, G. Photosynthetic symbionts and energy supply determine octocoral biodiversity in coral reefs. *Ecology* **2008**, *89*, 3163–3173. [CrossRef]

40. Kearse, M.; Moir, M.; Wilson, A.; Stones-havas, S.; Cheung, M.; Sturrock, S.; Buxton, S.; Cooper, A.; Markowitz, S.; Duran, C.; et al. Geneious Basic: An integrated and extendable desktop software platform for the organization and analysis of sequence data. *Bioinformatics* **2012**, *28*, 1647–1649. [CrossRef] [PubMed]

41. Hall, T.A. BioEdit: A user-friendly biological sequence alignment editor and analysis program for Windows 95/98/NT. *Nucleic Acids Symp. Ser.* **1999**, *41*, 95–98.

42. Larsson, A. AliView: A fast and lightweight alignment viewer and editor for large datasets. *Bioinformatics* **2014**, *30*, 3276–3278. [CrossRef]

43. Katoh, K.; Standley, D.M. MAFFT Multiple sequence alignment software version 7: Improvements in performance and usability. *Mol. Biol. Evol.* **2013**, *30*, 772–780. [CrossRef] [PubMed]

44. Ranwez, V.; Harispe, S.; Delsuc, F.; Douzery, E.J.P. MACSE: Multiple Alignment of Coding SEquences accounting for frameshifts and stop codons. *PLoS ONE* **2011**, *6*, e22594. [CrossRef]

45. Gaurav, V.; Lohman, D.J.; Meier, R. SequenceMatrix: Concatenation software for the fast assembly of multi-gene datasets with character set and codon information. *Cladistics* **2011**, *27*, 171–180. [CrossRef]

46. Stamatakis, A. RAxML version 8: A tool for phylogenetic analyses and post-analysis of large phylogenies. *Bioinformatics* **2014**, *30*, 1312–1313. [CrossRef]

47. Aberer, A.J.; Kobert, K.; Stamatakis, A. ExaBayes: Massively parallel Bayesian tree inference for the whole-genome era. *Mol. Biol. Evol.* **2014**, *31*, 2553–2556. [CrossRef] [PubMed]

48. Foster, P.G. Modeling compositional heterogeneity. *Syst. Biol.* **2014**, *53*, 485–495. [CrossRef] [PubMed]

49. Rambout, A. FigTree. 2014. Available online: tree.bio.ed.ac.uk/software/figtree/ (accessed on 17 April 2019).

50. Kumar, S.; Stecher, G.; Li, M.; Knyaz, C.; Tamura, K. MEGA X: Molecular evolutionary genetics analysis across computing platforms. *Mol. Biol. Evol.* **2018**, *35*, 1547–1549. [CrossRef] [PubMed]

51. McFadden, C.S.; Benayahu, Y.; Pante, E.; Thoma, J.N.; Nevarez, P.A.; France, S.C. Limitations of mitochondrial gene barcoding in Octocorallia. *Mol. Ecol. Resour.* **2011**, *11*, 19–31. [CrossRef] [PubMed]

52. Alderslade, P.; McFadden, C.S. Pinnule-less polyps: A new genus and new species of Indo-Pacific Clavulariidae and validation of the soft coral genus *Acrossota* and the family Acrossotidae (Coelenterata: Octocorallia). *Zootaxa* **2007**, *1400*, 27–44. [CrossRef]

53. LaJeunesse, T.C.; Parkinson, J.E.; Gabrielson, P.W.; Jeong, H.J.; Reimer, J.D.; Voolstra, C.R.; Santos, S.R. Systematic revision of Symbiodiniaceae highlights the antiquity and diversity of coral endosymbionts. *Curr. Biol.* **2018**, *28*. [CrossRef] [PubMed]

54. Benayahu, Y.; McFadden, C.S.; Shoham, E.; Van Ofwegen, L.P. Search for mesophotic octocorals (Cnidaria, Anthozoa) and their phylogeny: II. A new zooxanthellate species from Eilat, northern Red Sea. *ZooKeys* **2017**, *676*. [CrossRef]

55. LaJeunesse, T.C.; Thornhill, D.J. Improved resolution of reef-coral endosymbiont (*Symbiodinium*) species diversity, ecology, and evolution through *psbA* non-coding region genotyping. *PLoS ONE* **2011**, *6*, e29013. [CrossRef]

56. Alderslade, P.; McFadden, C.S. A new sclerite-free genus and species of Clavulariidae (Coelenterata: Octocorallia). *Zootaxa* **2011**, *3104*, 64–68. [CrossRef]

57. Benayahu, Y.; McFadden, C.S.; Shoham, E. Search for mesophotic octocorals (Cnidaria, Anthozoa) and their phylogeny: I. A new sclerite-free genus from Eilat, northern Red Sea. *ZooKeys* **2017**, *680*. [CrossRef]

58. Baldwin, C.C.; Pitassy, D.E.; Robertson, D.R. A new deep-reef scorpionfish (Teleostei, Scorpaenidae, Scorpaenodes) from the southern Caribbean with comments on depth distributions and relationships of western Atlantic members of the genus. *ZooKeys* **2016**, *606*, 141–158. [CrossRef] [PubMed]

59. Loya, Y.; Eyal, G.; Treibitz, T.; Lesser, M.P.; Appeldoorn, R. Theme section on mesophotic coral ecosystems: Advances in knowledge and future perspectives. *Coral Reefs* **2016**, *35*. [CrossRef]

 *diversity*

Article

Differential Occupation of Available Coral Hosts by Coral-Dwelling Damselfish (Pomacentridae) on Australia's Great Barrier Reef

Tory J Chase * and Mia O Hoogenboom

Marine Biology and Aquaculture Group, College of Science and Engineering, and ARC Centre of Excellence for Coral Reef Studies, James Cook University, Townsville QLD 4811, Australia; mia.hoogenboom1@jcu.edu.au
* Correspondence: tory.chase@my.jcu.edu.au

Received: 14 October 2019; Accepted: 8 November 2019; Published: 15 November 2019

Abstract: Associations between habitat-forming, branching scleractinian corals and damselfish have critical implications for the function and trophic dynamics of coral reef ecosystems. This study quantifies how different characteristics of reef habitat, and of coral morphology, determine whether fish occupy a coral colony. In situ surveys of aggregative damselfish–coral associations were conducted at 51 different sites distributed among 22 reefs spread along >1700 km of the Great Barrier Reef, to quantify interaction frequency over a large spatial scale. The prevalence of fish–coral associations between five damselfish (*Chromis viridis, Dascyllus aruanus, Dascyllus reticulatus, Pomacentrus amboinensis* and *Pomacentrus moluccensis*) and five coral species (*Acropora spathulata, Acropora intermedia, Pocillopora damicornis, Seriatopora hystrix,* and *Stylophora pistillata*) averaged ~30% across all corals, but ranged from <1% to 93% of small branching corals occupied at each site, depending on reef exposure levels and habitat. Surprisingly, coral cover was not correlated with coral occupancy, or total biomass of damselfish. Instead, the biomass of damselfish was two-fold greater on sheltered sites compared with exposed sites. Reef habitat type strongly governed these interactions with reef slope/base (25%) and shallow sand-patch habitats (38%) hosting a majority of aggregative damselfish-branching coral associations compared to reef flat (10%), crest (16%), and wall habitats (11%). Among the focal coral species, *Seriatopora hystrix* hosted the highest damselfish biomass (12.45 g per occupied colony) and *Acropora intermedia* the least (6.87 g per occupied colony). Analyses of local coral colony traits indicated that multiple factors governed colony usage, including spacing between colonies on the benthos, colony position, and colony branching patterns. Nevertheless, the morphological and habitat characteristics that determine whether or not a colony is occupied by fish varied among coral species. These findings illuminate the realized niche of one of the most important and abundant reef fish families and provide a context for understanding how fish–coral interactions influence coral population and community level processes.

Keywords: coral-fish association; symbiosis; habitat structure; prevalence; damselfish; coral reefs; biological interactions

---

1. Introduction

Scleractinian corals are the predominant habitat-forming organisms within coral reef ecosystems contributing to the (i) overall structure of reef habitats [1], (ii) co-existence and biodiversity of reef associated species [2–4], and (iii) providing critical microhabitats used by specialist species [5–8]. Consequently, the abundance of coral-dwelling and reef-associated species (e.g., crustaceans, sponges, bryozoans, fishes) is influenced by the abundance of habitat-forming corals [2,9], as well as by the structural complexity provided by coral-rich habitats [4,10–12], and the diversity of corals [13]. Importantly, high coral cover and habitat complexity moderate predation [14] and competition [15] among reef fish species. Meanwhile, fishes that have an intimate and obligate reliance on live corals for shelter (e.g., coral Gobiidae spp. [16], coral-dwelling Pomacentridae spp. [7]) or food (e.g., coral-feeding Chaetodontidae spp. [17]), often have specific preferences for select coral species which, themselves, might occur only in certain habitats (under certain environmental conditions or shelf positions [18]). Ultimately, corals might be a limiting resource that regulates the distribution and abundance of many reef fishes [6,19], depending on their specificity to particular coral species and their reliance on live coral habitats. Understanding this process requires intensive and broad-scale quantification of fish–coral interactions to distinguish effects of habitat types from the effects of coral cover.

The abundance of suitable coral, that enable long-term usage or residency of associated fauna across various life stages, is one of the most importance factors dictating damselfish presence [20,21], evident by the fact that abundances of fishes and motile invertebrates' abundances decline sharply following coral mortality [19,22]. Despite a strong dependence on corals by several fish families, not all coral colonies are occupied by fishes due to physical and behavioural limitations [7,21]. At a larger spatial scale of reefs and latitude, local availability of specific types of habitat determine spatial distribution patterns in habitat-specialized fish, (i.e., *Gobiodon* spp., [23,24]). However, to determine the extent to which the availability of specific coral habitats constrain the abundance of reef fishes, direct measurement of the abundance of fishes on individual coral colonies is required. Previous studies have linked variation in damselfish's abundance and diversity with habitat-related variation in the percentage cover or functional diversity of corals [18,25], but have not assessed whether and how features of coral colonies within habitats also influence fish abundance. Assessment of fish-coral interactions at the colony level is important, because this is the scale at which impacts of damselfish on corals are the most prevalent [26,27]. Services that fish provide to corals are often density-dependent (such as nutrient provision) and are heavily dependent on fish biomass [26,28]. Furthermore, understanding the spatial variation in coral-dwelling fish provides a context for understanding how these fish influence coral populations and communities, and how these mutualisms are likely to change during external disturbances and degradation.

Habitat type and colony morphology influence the suitability of coral hosts for nearly all types of coral-associated fauna, especially fishes, as observed for both Scleractinian and Alcyonacean corals [29–32]. This colony-scale association correlates with fish size [33], with how fish utilize the coral, and with fish diet preferences, and social and spatial niches [29,34,35]. Indeed, seascape and colony features strongly influence feeding behaviour, especially for zooplankton feeding damselfishes for which among-species partitioning of planktonic prey dictates how different fish species are distributed among reef zones [36–38]. Moreover, for other fish taxa, including Pomacentridae, Gobiidae, and Blennidae, fine-scale (1 to 10 cm^2) differences in the suitability of coral hosts depends on much more than just the coral taxonomic identity. For example, *Dascyllus marginatus* and other aggregating damselfish are more likely to inhabit colonies with fine branches, compared with lobed branching morphologies, as inter-branch space is a limiting determinant for these fishes [21,39]. Furthermore, seascape features such as distance from the reef's edge and water flow velocities govern species-specific patterns and biomass due to fishes' swimming ability, plankton/prey availability, trophic specializations [37], and colony arrangement requirements; this has been demonstrated in habitat-specialist/coral dwelling and non-coral-dwelling fishes [38,40–43]. However, the specific reef habitat characteristics, and with both within- and among-species variation in coral colony structures, that promote occupancy and residency by aggregative damselfish has yet to be fully understood, with most of this work to date focusing on Blennidae and Gobiidae that usually inhabit corals as individuals or pairs rather than in large aggregations [23]. Aggregative species are likely to utilize different resources and have different association levels compared with large transient and/or small solitary species [44].

This research explores the large-scale spatial variation in occupancy rates and biomass of coral-dwelling damselfish among predominant habitat-forming scleractinian coral species, and assesses specific habitat and colony features that influence whether or not individual coral colonies are used by coral-dwelling, planktivorous damselfish. The prevalence of fish-coral interactions is examined for five damselfish (*Chromis viridis, Dascyllus aruanus, Dascyllus reticulatus, Pomacentrus amboinensis,* and *Pomacentrus moluccensis*) on five coral species (*Acropora spathulata, Acropora intermedia, Pocillopora damicornis, Seriatopora hystrix,* and *Stylophora pistillata*). These corals are frequently occupied by coral-dwelling damselfish [5,21,45–47]. Previous research demonstrates that these select coral and fish species can account for >70% of non-cryptic fish-coral interactions within the Great Barrier Reef [21,47] and play important roles in assimilating energy and nutrients from plankton into the reef food web. Each of these Pomacentrid sp. have been documented to be 'coral-dwelling' with a home range of a single coral or similar <2 m structure [18,27,39,48,49]. Finally, the focal fish species represent important prey for meso- and top predators [18] and are therefore important in reef trophic dynamics. Multiple coral colony traits were measured in situ as these traits are hypothesized to correlate with patterns of fish occupation and biomass. This study increases the number studies that have quantified broad-scale abundance of damselfish associated with different colony morphologies. We investigate the distribution of these coexisting damselfish within and among reef zones to (a) determine if suitable coral habitat governs patterns in damselfish's distribution and abundance (large scale, >10 m, based on variation in coral cover among reefs), and (b) quantify variations in fish biomass within and among coral colony species (small scale >1 m, based on observations of individual coral colonies). Evaluating the multiscale spatial variation of fish-coral interactions provides insight into fishes' effects on coral health, and context for predicting the functioning of interspecific and symbiotic associations during global environmental change.

2. Materials and Methods

2.1. Study Sites and Surveys

This study was conducted in March–November 2016, in the northern Great Barrier Reef (GBR), Australia. Surveys were conducted at 51 study sites across 20 different reefs (Figure 1) including the far northern sector (*n* = 11 sites), the northern sector (*n* = 24) including Lizard Island sites (*n* = 16 sites), the central sector (*n* = 13 sites) and the southern sector near One Tree Island (*n* = 3 sites, see Table 1).

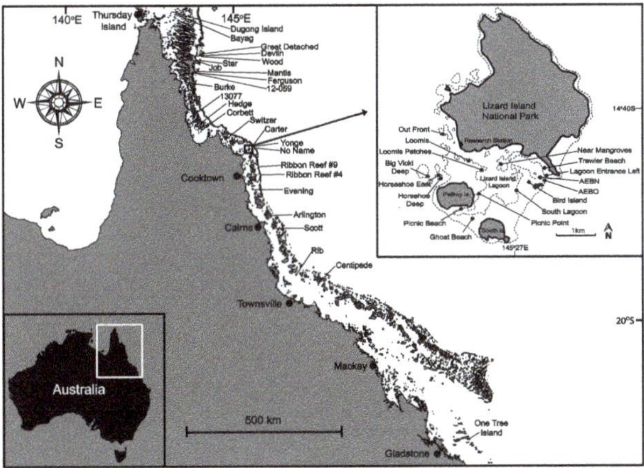

Figure 1. Location of study reefs (for reef seascape surveys and colony level surveys) along the Great Barrier Reef (GBR, 51 sites spread among 22 different reefs), spanning >1700 km (map modified from [50]) with Lizard Island subset including 16 sites, surveyed between February and November 2016. Some reefs contained more than one transect. Map template is provided by Geoscience Australia under a Creative Commons Attribution 4.0 International License.

Table 1. Dataset summaries detailing the (a) observed site features quantified for 51 sites, and the individual coral (b) colony orientation, and (c) colony structure for 226 individual colonies with and without resident damselfish (occupancy and biomass as independent variables) on mid-shelf and offshore GBR reefs. Coral-fish interactions were focused on five focal fish species (*Chromis viridis, Dascyllus aruanus, Dascyllus reticulatus, Pomacentrus amboinensis,* and *Pomacentrus moluccensis*) occupying five common branching morphologies (*Acropora intermedia, Acropora spathulata, Pocillopora damicornis, Seriatopora hystrix,* and *Stylophora pistillata*).

(a) Reef Seascape Surveys:	Description of Data
Sampling: 50 × 5 m belt transects n =3097 colonies on 51 sites spread among 22 reefs, including sheltered (n = 28 sites) and exposed (n = 23 sites) sites, and sand patches (n = 11 sites), reef flat (n = 5 sites), reef crest (n = 12 sites), reef wall (n = 5 sites), and reef slope (n =18 sites) habitats.	
Latitude	Sites grouped by latitude: Far north (12.3° S to 10.5° S), North (12.3° S), Central (14.7° S to 18.9° S) and South (One Tree Island, 23.5° S)
Aspect	Exposed locations (high water flow, mainly eastern side of GBR reefs) vs sheltered locations (low water flow, mainly western side of GBR reefs)
Habitat Zone	Lagoon sand patch, reef flat, reef crest, reef wall, and reef slope/base
Benthic cover [51]	Percentage cover measured on four 10 m line intercept transects at each site using the following categories: *Isopora, Montipora,* tabular *Acropora,* staghorn *Acropora,* other *Acropora, Pocillopora damicornis, Seriatopora, Stylophora,* other pocilloporids, Mussidae, Faviidae, Poritidae, other scleractinians, soft corals, and other sessile fauna
Coral Species	*Acropora intermedia, Acropora spathulata, Pocillopora damicornis, Seriatopora hystrix,* and *Stylophora pistillata*
(b) Colony orientation: *Sampling: Digital photos of n = 226 colonies on 15 sites on 11 exposed and sheltered reefs*	
Colony Orientation [52] (Position of Coral on Substratum)	Crevice—colony grew within a crack in the reef matrix; Open—colony is on flat reef benthos without any obvious shading by competitors; Sand—colony grew above a sand patch; Underhang—colony was shaded by reef matrix or other colonies
(c) Colony structure: *Sampling: Digital photos of n = 226 colonies on 15 sites on 11 exposed and sheltered reefs*	
Colony Diameter	Average of the longest colony diameter dimension and the diameter perpendicular to that dimension
Planar Area	Digitally traced along colony perimeter
Colony Height	Distance from top of coral colony to substrate
Branch Width	Average width of branches (n = 5 branches colony^{-1}) located throughout the colony
Branch Spacing	Average distance between branches (n = 5 branches colony^{-1}) located throughout the colony
Isolation	Distance to nearest branching, columnar, tabular, or foliaceous colony

Along latitudes spanning >10°, 1–3 transects per site were compared to quantify occupancy and resident damselfish's biomass. Sites were either sheltered or exposed; western facing aspects (sheltered sites, often with sandy lagoons) receive less exposure of wave energy and weather, compared with eastern facing aspects (exposed) sites on the GBR, due to the geomorphology of the surveyed mid-shelf and off-shore reefs [53,54]. Transects were located within different habitat zones (sand patches, flat, crest, wall (distinguished from slope by approximately vertical relief of the substratum), slope/base (gentle gradient or approximately flat), at different distances from shore (mid-shelf and off-shore reefs), and at varying depths (0–14 m, standardized to Lowest Astronomical Tide (LAT)). Herein, occupancy is described as a colony being used as the sole site of shelter/habitat within a damselfish's territory or home range [21,22,55]. Surveys focused on five species of damselfish (*Chromis viridis*, *Dascyllus aruanus*, *Dascyllus reticulatus*, *Pomacentrus amboinensis*, and *Pomacentrus moluccensis*) and five species of branching corals (*Acropora intermedia*, *Acropora spathulata*, *Pocillopora damicornis*, *Seriatopora hystrix*, and *Stylophora pistillata*, see Figure S1 in Supplementary Materials). The host corals were selected for their abundance on the GBR, while also displaying differences in morphology, and particularly, branch spacing patterns due to the hypothesized role of branch spacing in determining colony occupancy [9,56].

At each site, the abundance and occupation of colonies (20–100 cm in diameter) of each study species (*Acropora intermedia*, *Acropora spathulata*, *Pocillopora damicornis*, *Seriatopora hystrix*, and *Stylophora pistillata*) were recorded along a 50 m × 5 m belt transect (total area of 250 m^2) by scuba diving. We also recorded the size and abundance of focal fish species (*Chromis viridis*, *Dascyllus aruanus*, *Dascyllus reticulatus*, *Pomacentrus amboinensis*, and *Pomacentrus moluccensis*) within each colony through a visual census during scuba diving. Along each transect, each colony was slowly approached and observed for at least 30 s to determine damselfish species presence, size, and abundance for biomass estimates. For consistency, all coral and fish observations were performed by the same observer during daylight hours (between 8:00 and 18:00 h). In addition, four replicate 10 m line intercept transects were completed at each site to measure total coral cover (of all corals not just the 5 focal species [51,57]).

To assess whether occupation of focal coral species by the specific damselfish was influenced by intrinsic or extrinsic factors, we measured a series of colony attributes for a subset of colonies ($n = 226$) at 15 different sites. These colonies were located on 11 exposed and sheltered reefs, spanning habitats at a depth range of 0–13 m, positioned in the Far North, North, Central, and Southern GBR regions as described above (see Tables S3–S5 in Supplementary Materials for details). Colony position was categorized as being either within a crevice, on an overhang, on open carbonate pavement, or on sand [52]. Colony structure traits measured included: colony size (colony diameter, planar area, and colony height), distance from nearby corals (isolation), and branch dimensions (i.e., inter-branch spacing and average branch width [47,58], see Table 1 and Figure S2 in Supplementary Materials). Branch spacing and branch width were averaged for five measurements around each colony, with all branch measurements taken at ~15 mm from the branch tip, while colony isolation being measured as the distance to the closest habitat providing coral (i.e., branching or other complex morphology colonies). For colonies with resident fishes ($n = 142$), the numbers of all fishes on each focal colony were recorded, and all fish were placed into general standard-length size classes of small, medium, and large, for each species respectively. Size class data were subsequently converted into biomass estimates, based on published length/weight relationships generated from damselfish [27,47], where damselfish were collected using hand-nets and a liquid anesthetic (a diluted solution of clove oil, ethanol, and seawater [27,59,60]). Surveys focused on ecologically important damselfish's occupancy and biomass patterns rather than fish numbers, as biomass has been directly linked to fish-derived services and benefits for corals [26,61]. For the purposes of these surveys, fish biomass summarizes both fish numbers and size, and the analysis did not delineate which of these components contribute more to biomass levels. Additional details of transects, sites and colonies are provided in Table 1 and Supplementary Tables S3–S5.

2.2. Data Analysis

2.2.1. Reefscape Prevalence of Fish–Coral Interactions

At the reef seascape level, the proportion of colonies occupied by fish (all damselfish and coral species pooled, as the independent variable) varied on each transect, and was analysed using a full additive beta regression model with latitude, aspect (exposure level), habitat, and coral cover as fixed dependent factors, and reef as a random factor. A beta regression was deemed appropriate, as it includes a logit transformation which is necessary for proportional data [62,63]. The appropriateness of the models selected were confirmed by assessing quantile, or Q-Q plots for normality and residual plots for homogeneity of variance and linearity, as well as calculations of dispersion. Additive models (latitude + aspect + habitat + coral cover) were used due to the non-factorial nature of the dataset wherein not all habitats and aspects could be sampled at each latitude.

A linear mixed-effect model (LME) was used to analyse effects of latitude, aspect, habitat, and coral cover, on total biomass of focal damselfish species (grams per 250 m^2), log +1 transformed, recorded on each transect. The fish biomass LME was fitted using maximum likelihood [64]. Model selection, based on Akaike information criteria (AIC) values, was implemented to determine the importance of latitude, aspect, habitat, and coral cover as predictors of fish biomass (following [65,66], see below) and assumptions for model validity were checked through Q-Q plots (normality) and residual plots (homogeneity of variance and linearity), as well as calculations of dispersion. In addition, the multi-model interference R package *MuMIn*, was used to perform model selection on proportion of colonies occupied and total biomass models based on model weights derived from AICc. *MuMIn* allows for an estimate of the variance explained by all factors included in the model (*R package MuMIn*, [67,68]). A ranking of the possible models to identify the contribution/importance of each variable as well as the number of models in which each variable was completed (function "dredge" in R package *MuMIn*).

To compare differences in total occupancy (only occupied colonies, $n = 898$) among each of the five coral species, binomial generalized linear models (GLMs) with Tukey's honestly significant difference (HSD) post hoc (with Bonferroni adjusted p-values) were used. Total damselfish's biomass was analyzed using a Gaussian GLM with Tukey's HSD post hoc. Separate Kruskal–Wallis rank sum tests were performed for each damselfish species to analyse whether coral species identity (independent variable) affected the biomass of different species of resident damselfish (dependent variable) on these 898 occupied colonies. Kruskal–Wallis tests were deemed appropriate as fish biomass data did not meet assumptions of homogeneity of variance and normality. Dunn tests were used for multiple post hoc comparisons between species due to unequal sample sizes, and p-values were adjusted with the Benjamini–Hochberg method to decrease type I error; the Benjamini–Hochberg method is a more powerful method than the Bonferroni correction to control the false discovery rate [69] and frequently used with the Kruskal–Wallis test.

2.2.2. Effects of Colony Position and Structure on Damselfish's Occupancy

To compare how colony position and structure impacted occupation and biomass for a subset of colonies, principle component analyses (PCAs) were used to evaluate overall differences in colony morphology between corals with ($n = 142$) and without fish ($n = 84$), both with data pooled over all corals ($n = 226$), and separately for each coral species (using the colony level dataset). These different analyses were conducted to assess whether there were particular colony structure traits that influenced fish presence overall, and whether such features were consistent among coral species. PCAs were deemed appropriate due to the multivariate nature of the data with variables (e.g., branch width and branch spacing) that were likely to be correlated with each other. The PCA ordinated colonies were based on the standardized correlation matrix between colony attributes using the R function princomp [70,71]. Subsequently, the principle component (PC) 1 and 2 scores of each colony were used to represent the overall variation in colony morphology in subsequent linear models (LM) of fish occupation (presence/absence). To further differentiate occupancy patterns between the colony position, a binomial GLM was used with a Tukey's HSD post hoc to assess between factor level differences. A lognormal linear model was used to quantify total damselfish's biomass (only occupied colonies) with regards to colony position, again with Tukey's HSD post hoc comparisons.

Similarity percentage analysis (SIMPER [71–73]) was used to determine which coral structure traits (colony diameter, planar area, colony height, branch spacing, branch width, and isolation) contributed the most to the differences among corals with and without fish. This analysis compared the importance of these structural traits for all coral species pooled and pooled across the different species of fish occupying these corals. The SIMPER analysis was performed on the PCA standardized data to assess which structure traits were driving the differences (by individual coral species and species pooled) and ranked in order according to their contribution (% or importance ranking). This similarity percentage is based on the decomposition of Bray-Curtis dissimilarity index, giving the overall contribution of individual structure traits.

2.2.3. Effects of Colony Position and Structure on the Biomass of Damselfish

Total biomass of damselfish on colonies located in different reef microhabitats (colony orientation, Table 1) were analysed with lognormal linear models. Model fit was assessed using residual plots, all of which were satisfactory (normal and homogenous). As total pooled damselfishes' biomass is a continuous variable, a series of linear models per individual coral species and for all colonies pooled were completed to determine if total damselfishes' biomass (dependent variable) varied with the two most important structure traits (independent variables) from the SIMPER of colony structure occupancy.

All data analyses were performed in the statistical software R [74] using the *betareg* [62], *multcomp* [75], *lsmeans* [76], *simper* function in *vegan*, and *MuMIn* [67,68] packages. Full datasets are available at [77].

3. Results

3.1. Range of Damselfish's Occupations across the Great Barrier Reef (GBR)

In transect-based surveys, a total of 5154 damselfish of the five focal species (*Chromis viridis, Dascyllus aruanus, Dascyllus reticulatus, Pomacentrus amboinensis*, and *Pomacentrus moluccensis*) were counted on 3034 coral colonies of the five focal species (*Acropora intermedia, Acropora spathulata, Pocillopora damicornis, Seriatopora hystrix*, and *Stylophora pistillata*) on 51 transects (with combined sample area of 12,750 m^2). Overall, 30% of colonies were occupied by one or more of the focal damselfish species (898 out of 3034, all transects pooled). Single-species groups of *Pomacentrus moluccensis* or *Dascyllus aruanus* were recorded on 80% of occupied colonies. *P. moluccensis* were prevalent (in terms of coral host occupancy) in all habitats, while *Chromis* and *Dascyllus* species almost exclusively inhabited corals on sand-patch and slope habitats (Figure S2 in Supplementary Materials).

Occupancy varied with aspect and habitats, with values ranging from 0% at exposed, flat and crest habitat zones, up to 93% at sheltered sand-patch habitats. In the full model, habitat (1) and aspect (2) were the most important variables in predicting fish occupancy (Table S1 in Supplementary Materials). In general, occupancy levels were higher in western aspect/sheltered sites locations (including lagoons) than eastern aspect/exposed sites (betareg (logit): $p = 0.002$), and highest numbers were observed in sand patches and slope habitats ($p = 0.016$, Figure 2b). Latitude ($p = 0.051$), and coral cover ($p = 0.735$) were not significant predictors of the proportion of colonies occupied (Figure 2a,b).

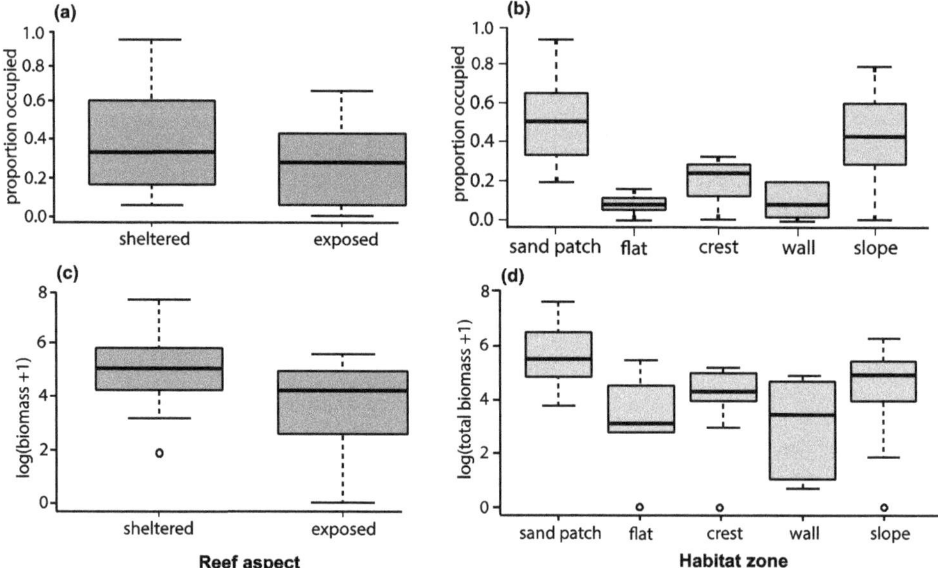

Figure 2. Boxplots (horizontal lines show median; boxes indicate 25th and 75th percentiles; vertical dotted lines show range; data points show outliers) of colonies occupied (reef seascape) (**a**,**b**) and damselfish's biomass (log +1) abundance (**c**,**d**) on five species of branching coral (*Acropora intermedia, Acropora spathulata, Pocillopora damicornis, Seriatopora hystrix,* and *Stylophora pistillata*) in relation to aspect category (exposed or sheltered) and reef habitat (sand patches, flat, crest, wall, and slope/base).

Additionally, occupancy also varied with coral species (binomial GLM, significant effect of species, $p < 0.05$). Both *P. damicornis* (34% occupancy) and *Stylophora pistillata* (33% occupancy) had the highest average occupancy, when compared with *Acropora spathulata* (30%), *S. hystrix* (23%), and *Acropora intermedia* (22%) (see Table 2 for post hoc comparisons and Table S2 for the binomial GLM output). These damselfish species-specific occupancy patterns translated into different damselfish diversity and biomass on each coral species (Tables S3–S5 in Supplementary Materials); for instance, *Acropora intermedia, Pocillopora damicornis,* and *Stylophora pistillata* hosted mainly *Dascyllus aruanus* and *Pomacentrus moluccensis* aggregations, while *Acropora spathulata* hosted *Chromis viridis* and *Pomacentrus moluccensis* heterospecific groups.

Table 2. Multiple comparisons of coral-species, with *p*-values, (Tukey's honestly significant difference (HSD) post hoc) based on a binomial generalized linear model of colony occupancy with damselfish species pooled (reef seascape dataset): colony occupancy (dependent) and colony species (independent variable). Significant *p*-values are in bold.

Comparison	*p*-Value
A. intermedia – A. spathulata	0.5089
A. intermedia – P. damicornis	**0.0050**
A. intermedia – S. hystrix	0.9996
A. intermedia – S. pistillata	**0.0131**
A. spathulata – P. damicornis	0.8963
A. spathulata – S. hystrix	0.4492
A. spathulata – S. pistillata	0.9588
P. damicornis – S. hystrix	**<0.001**

3.2. Patterns of Damselfish Biomass across Reefs on Occupied Colonies

An average of six damselfish were present on each occupied colony. *Pomacentus amboinensis* was the most prevalent damselfish species on the coral colonies considered during this study, present on nearly half of all occupied coral colonies (~2.3 *Pomacentus moluccensis* colony^{-1}), and accounting for ~45% of all damselfish's biomass on coral hosts (Table 3, Figure 3, and Tables S3–S5 and Figure S3 in Supplementary Materials). *Dascyllus aruanus* was the second most abundant with an average 1.8 fish per occupied colony^{-1} and the other three species were present at considerably lower abundance (*Chromis viridis*: 0.8 fish occupied colony^{-1}, *Dascyllus reticulatus* 0.2 fish occupied colony^{-1}, and *Pomacentrus moluccensis*: 0.5 fish occupied colony^{-1}).

Damselfish's biomass was broadly similar to occupancy patterns, displaying significant differences in biomass per 250 m^2 depending on aspect (LME (log+1), aspect, $\chi^2 = 6.88$, $p = 0.008$, Figure 2c,d). Sheltered sites had three-fold higher biomass (250 ± 71 g 250 m^{-2} for all colonies per transect) than exposed sites (86.7 ± 17 g 250 m^{-2}). Biomass per 250 m^2 also varied by habitat zone (LME (log+1), habitat, $\chi^2 = 9.54$ $p = 0.0489$) with the highest biomass in sand patches (404.9 ± 166 g 250 m^{-2}) and slope habitats (161.7 ± 33 g 250 m^{-2}), and lowest biomass on wall habitats (70.1 ± 42 g 250 m^{-2}). Again, latitude (LME (log+1) $\chi^2 = 2.81$, $p = 0.42$) and coral cover ($\chi^2 = 0.109$, p = 0.740) were not significant predictors of total fish biomass per transect. In the full model, aspect (1) and habitat (2) were the most important variables in predicting fish occupancy (Table S1 in Supplementary Materials).

Damselfish's biomass per occupied colony ranged from 1.3 g (a single *Pomacentrus amboinensis*) to 120 g (a school of ~100 *Chromis viridis* or a large aggregation of ~30 *Dascyllus aruanus*). Among the five fish species, *Pomacentrus moluccensis* exhibited the most consistent and broadest distribution being present in high biomass in every habitat zone. *Seriatopora hystrix* coral colonies hosted the highest fish biomass per occupied colony (12.45 g ± 1.33), with *Acropora intermedia* having the lowest biomass per occupied colony (6.87 g ± 1.33). As a result, total damselfish's biomass was significantly different among occupied coral species (GLM: $p = 0.012$, see Supplementary Table S6 for post hoc comparisons). When data were analysed by fish species, the biomass of each damselfish species significantly varied among host coral species (see Table S7 in Supplementary Materials for post hoc comparisons), except for *Chromis virdis* (Kruskal–Wallis: $\chi^2 = 9.104$, df = 4, $p = 0.0586$). *Seriatopora hystrix* and *Pocillopora damicornis* colonies were favoured by *Dascyllus aruanus* ($\chi^2 = 45.304$, df = 4, $p < 0.001$) and *Dascyllus reticulatus* ($\chi^2 = 29.962$, df = 4, $p < 0.001$). *Acropora spathulata* and *Stylophora pistillata* colonies were favoured by *Pomacentrus amboinensis* ($\chi^2 = 11.715$, df = 4, $p = 0.019$) and *Pomacentrus moluccensis* ($\chi^2 = 29.962$, df = 4, $p < 0.001$).

Table 3. Average biomass estimates (mean ± SE) for each damselfish species (*Chromis viridis*, *Dascyllus aruanus*, *Dascyllus reticulatus*, *Pomacentrus amboinensis*, and *Pomacentrus moluccensis*) on each coral species (*Acropora intermedia*, *Acropora spathulata*, *Pocillopora damicornis*, *Seriatopora hystrix*, and *Stylophora pistillata*) on occupied colonies in sheltered and exposed aspect sites.

Aspect	Coral Species	*n*	Average Biomass (g) per Coral Species per Site Aspect				
			C. viridis	*D. aruanus*	*D. reticulatus*	*P. amboinensis*	*P. moluccensis*
Sheltered	*A. intermedia*	38	0.92 ± 0.79	2.21 ± 0.96	0.94 ± 0.89	1.06 ± 0.31	2.80 ± 0.54
	A. spathulata	30	4.66 ± 2.03	0.60 ± 0.48	0.17 ± 0.17	0.09 ± 0.06	5.70 ± 0.96
	P. damicornis	233	0.52 ± 0.22	5.00 ± 0.79	0.26 ± 0.14	0.84 ± 0.12	3.89 ± 0.34
	S. hystrix	147	2.63 ± 1.10	8.49 ± 1.15	0.00 ± 0.00	0.63 ± 0.11	2.47 ± 0.34
	S. pistillata	179	0.22 ± 0.02	4.72 ± 0.64	0.36 ± 0.12	0.50 ± 0.99	4.74 ± 0.51
Exposed	*A. intermedia*	16	0.19 ± 0.19	0.00 ± 0.00	0.00 ± 0.00	0.24 ± 0.13	3.91 ± 1.06
	A. spathulata	6	0.00 ± 0.00	0.00 ± 0.00	0.00 ± 0.00	0.00 ± 0.00	9.18 ± 3.31
	P. damicornis	86	0.24 ± 0.23	0.00 ± 0.00	1.52 ± 1.03	0.48 ± 0.16	6.45 ± 1.00
	S. hystrix	42	0.00 ± 0.00	0.00 ± 0.00	0.04 ± 0.04	0.84 ± 0.22	5.38 ± 0.89
	S. pistillata	121	0.00 ± 0.00	0.18 ± 0.12	0.22 ± 0.11	0.75 ± 0.16	5.79 ± 0.50

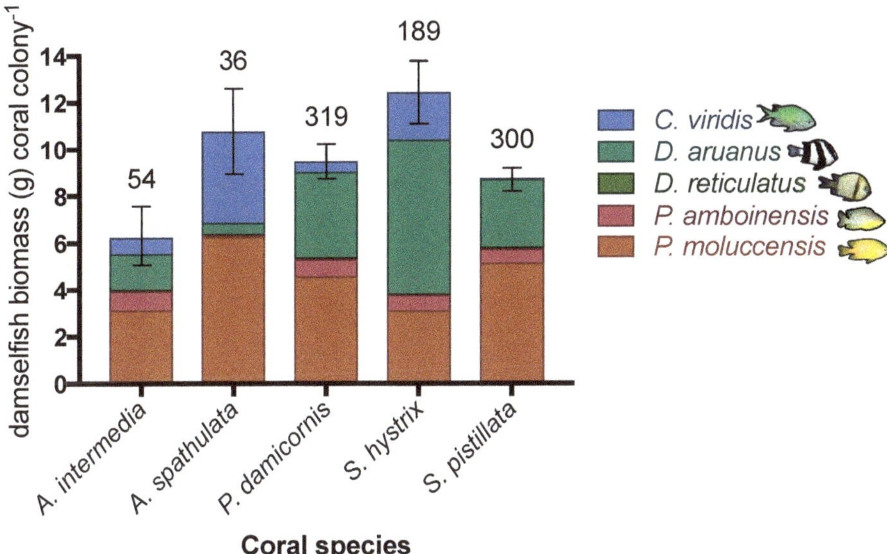

Figure 3. Mean biomass per fish species (g ± standard error (SE) of total biomass of damselfish) per coral species (*Acropora intermedia, Acropora spathulata, Pocillopora damicornis, Seriatopora hystrix,* and *Stylophora pistillata*) for all occupied colonies (*n* = 898) for 5154 fish (*Chromis viridis, Dascyllus aruanus, Dascyllus reticulatus, Pomacentrus amboinensis,* and *Pomacentrus moluccensis*) at 51 sites. Coral sample sizes per species are displayed above the bars. Note the collapse of *Dascyllus reticulatus* and *Pomacentrus amboinensis* sub-bars for the *Acropora spathulata* coral bar, and again for *Dascyllus reticulatus* on the *Seriatopora hystrix* bar, indicating very low biomass values for these fish species on these corals. Further damselfish species-specific and coral species-specific average biomass (±SE) per site aspect, and habitat are displayed in Supplementary Tables S3–S5).

3.3. Colony Orientation as a Determinant of Damselfish's Occupation and Group Biomass

Higher coral occupancy was observed on corals located in reef microhabitats that were either open carbonate pavement or open sandy substratum habitats (LM: (open) $p = 0.0068$) and (sand) $p < 0.0001$, see Table S8 for post hoc comparisons in Supplementary Materials). Similarly, total damselfish's biomass on occupied colonies (all fish and coral species pooled), averaged 15.3 g ± 2.4 on sand, and 11.4 g ± 1.8 on open colonies; values that were three- to four-fold higher than observed on colonies in underhang (4.9 g ± 0.8) and crevice (3.8 g ± 0.7) colony orientations (LM: $F_{3132} = 5.387$, $p < 0.001$, see Table S9 for post hoc comparisons in Supplementary Materials).

3.4. Colony Structure as a Determinant of Damselfish's Occupation and Biomass

The PCAs of colony attributes (based on the specific subset of corals and study locations where these attributes were measured) of the five coral-dwelling damselfish (species pooled), revealed distinctive groupings of colonies with and without fish both when data were pooled across coral species, and when analysed separately for each coral species. The first two principal components explained 70% of variance for all colonies pooled (Table S10 and Figure S4 in Supplementary Materials), and between 55% and 77% of variance in colony structure when coral species were analysed individually. Overall, colonies (pooled over species) that were occupied by fish had considerably lower PC1 scores than colonies without fish, and lower PC2 scores (Table S6). In this analysis, PC1 scores were associated with variation in colony diameter and planar areas (dictated by *Acropora intermedia* colonies), and PC2 scores were driven by branch spacing and colony isolation. When coral colonies were analysed separately by individual species (Supplementary Table S11) isolation was the most influential colony variable for all coral species, with branch spacing and planar area as secondary variables.

Total damselfish's biomass per colony followed similar trends with fish occupancy (Table 4 and Table S12 in Supplementary Materials), with isolation and colony height as the most influential colony-structure variables for five of the six coral species, and all structure traits were significant except for branch width, which when analysed individually by species, was only important for *Stylophora pistillata*. Branch spacing, colony diameter, and planar colony area were significant for three coral species. Branch width was only important for predicting fish biomass present on *Stylophora pistillata* colonies (Table 4).

Table 4. Series of linear models illustrating variation in total biomass of damselfish in small branching coral colonies (*Acropora intermedia*, *Acropora spathulata*, *Pocillopora damicornis*, *Seriatopora hystrix*, and *Stylophora pistillata*), by damselfish (*Chromis viridis*, *Dascyllus aruanus*, *Dascyllus reticulatus*, *Pomacentrus amboinensis*, and *Pomacentrus. moluccensis*) for six fine-scale indicators of colony attributes (colony level dataset). The first two traits, colony isolation and branch spacing (shaded), had the highest importance for determining colony occupation. Significant *p*-values are in bold.

Coral Species	*n*	Isolation	Branch Spacing	Colony Diameter	Planar Area	Colony Height	Branch Width
A. intermedia	22	**<0.001**	0.527	**0.019**	**<0.001**	0.068	0.185
A. spathulata	32	0.099	0.357	**0.020**	**<0.001**	**0.008**	0.416
P. damicornis	66	**0.014**	**0.002**	**<0.001**	**<0.001**	**<0.001**	0.423
S. hystrix	44	**0.002**	**<0.001**	0.304	0.809	**0.018**	0.198
S. pistillata	62	**<0.001**	**<0.001**	0.734	0.357	**0.015**	**<0.001**
All colonies	226	**<0.001**	0.007	**<0.001**	**<0.001**	**<0.001**	0.176

4. Discussion

This research demonstrates substantial variation in the occupancy rates of small-branching coral hosts by five species of coral-dwelling damselfish, with between 0%–93% of coral colonies being occupied per transect, depending on reef habitat zone and exposure. Within habitats, small-scale differences in the morphology and position of coral colonies also contributed to occupancy and biomass of fishes. Previous studies have suggested that variation in coral colony structure and health are likely to play important roles in determining the population dynamics of coral-associated fishes and invertebrates [78,79], as well as the persistence of these fish assemblages. This study provides new insight into the factors that control the presence and abundance of individual symbiotic damselfish species (and associated group biomass, distribution across parts of the GBR) and provides context for understanding the potential impacts of aggregating damselfish on complex networks of reef species and reef ecosystem function.

The overall rates of occupancy reported in this study (30%) were aspect- and habitat zone-specific, demonstrating patterns of both high occupancy and high biomass on patchy sheltered aspect sites and significantly lower values on continuous, exposed aspect sites. At the transect level, physical conditions of these habitats are congruent with many of the environmental gradients (i.e., water-flow) and niche partitioning requirements that structure damselfish populations [7,18,35,80]. These results suggest that generalist damselfish species may be better able to utilise corals as habitat in high-flow environments than other species that are limited to specific coral species which may be more prevalent in sheltered areas [21,22,41]. For instance, *Pomacentrus moluccensis* was the most prevalent damselfish species recorded and contributed disproportionally to the fish biomass present on occupied colonies on exposed sites. Consequently, damselfish inhabiting branching corals in exposed or deeper locations may be generalists zooplankton/omnivore feeders with the ability to take advantage of pelagic subsidies [81] rather than more specialized feeders [34,36,37,82]. While most coral-dwelling damselfish are found in sheltered habitats (i.e., flow $< 21.2\,\mathrm{cm\,s^{-1}}$), the body shape and fin morphology of *Pomacentrus moluccensis* may make them more adapted to higher current velocities, while *Dascyllus aruanus* may be more suited to lower currents [40,41] and was present on only 14% of damselfish-occupied corals, nearly all on sheltered sites. As an omnivorous bentho-pelagic feeder, *Dascyllus aruanus* can consume zooplankton and algae in equal proportions [34,38] which may partially explain its high abundance in slope/base habitats. Environmental factors such as water temperature, salinity, predators, conspecifics, and prey availability are also likely to influence the distribution and abundance of damselfish, independent of the abundance or availability of suitable coral hosts [34,37,83]. Although structural complexity and subsequent coral cover are often positively associated with fish biodiversity [1,4], results of this study showed that these two variables did not predict occupancy or biomass of coral-dwelling damselfish that closely associate with corals, consistent with previous studies [7,9,25,84–86]. Furthermore, latitude did not significantly affect colony occupancy or biomass; consistent with other studies reporting distribution and abundance of planktivorous damselfish along the Great Barrier Reef [18].

The coral species considered within this study (*Acropora intermedia*, *Acropora spathulata*, *Pocillopora damicornis*, *Seriatopora hystrix*, and *Stylophora pistillata*) are among the preferred coral hosts for coral-dwelling damselfish [21,22,46], yet 68% of colonies were unoccupied. This suggests that either abundance of these damselfish is not limited by coral host availability [87,88], or that there are colony attributes beyond species identity that determine their suitability as host corals [78,89]. This research reveals a suite of factors at small scale (<1 m) that influence occupation rates, including colony height, and colony position on the benthos, as well as the distance to other potential host corals. These attributes do not necessarily distinguish suitable versus unsuitable microhabitats, but given the choice of host corals, it would be expected that damselfish would select hosts that maximize individual fitness. Colonies with more elevated growth forms, raised above the seafloor, may also enhance fishes' abilities to stay higher in the water column, containing more enriched plankton, yet still close to refuge [90,91]. Elevated (high overall height) and isolated colonies, often in open position or on sandy substrates, allow for feeding with reduced danger due to visibility and enhanced colony structure complexity for refuge. Furthermore, damselfish species may respond differently to different species and morphologies of corals, with colony structure likely being more important to small-bodied fish [29,30]. For instance, *Dascyllus aruanus* prefer colonies with medium open-branch spacing (i.e., *Pocillopora damicornis* over *Acropora intermedia* or *Acropora spathulata*), while *Pomacentrus moluccensis* showed more equivalent abundance on all colony species. Branch spacing of corals limits occupancy only in tighter branching species (*Acropora spathulata*, *Pocillopora damicornis*, and *Seriatopora hystrix*) and may lead to variations in species interactions [92] between damselfish with their competitors and/or predators, and services (i.e., nutrient retention [26]).

Colony isolation was consistently the most important attribute predicting the presence and biomass of damselfish. Many damselfish species exhibit 'clumped' or 'patchy' distributions, leading to increased fish–coral interactions with increased fish abundance [47]. Sand patch and slope/base habitats, often categorized as edge habitats [43] with lower coral cover, host more fish–coral interactions and allow for more 'open' colonies, rather than nested corals along continuous reefs [39,93]. The isolation and spacing of colonies occupied may allow for: (a) continual use and residency by fish (i.e., distance to nearest available habitat is beyond the fish's home range); (b) increased impacts of association defense and reduction of fish predation [9,94–96]; (c) access to plankton resources and reduced competition [9,97–99]; and (d) larger borders with sandy substrates as an alternative foraging substrate [100]. Competition between damselfish species is also responsible for the ecological partitioning of these species along gradients [98], leading to differential use and fish-derived benefits to coral hosts [9,101]. Many of these factors may enhance the survival of the coral holobiont in select habitats.

Coral occupancy of 30% may be an underestimate, as it excludes additional common fish families that can inhabit coral colonies (i.e., Apogonidae, Gobiidae, Haemulidae), and coral sizes (>100 cm), and coral species (i.e., *Porites* and *Echinopora*). While damselfish are present in many coral reef habitats [18], fish–coral interactions may vary in the sign or magnitude of the effect on their coral host [47,92], with sand patch, and slope/base zones acting as small-scale interaction hotspots with high occupancy and biomass patterns. These hotspots, areas of high localized nutrient production by fishes [28,102] are generally infrequent across seascapes. Nutrient subsidy, along with other fish-derived services like increased photosynthesis [103], colony growth [28], bleaching susceptibility [27] sediment removal [101], may be density-dependent (i.e., >15 g seen in studies focusing on larger-bodied or more abundant fish species, see [26,104,105]) and fish-species dependent. With 68% of corals vacant, it is clear that many colonies do not receive potential beneficial effects of resident damselfish. However, certain provided benefits, such as increased oxygen input [103] or nutrients [26], may be more necessary within specific habitats (i.e., deeper sand patch and slope habitats) or under specific environmental conditions (i.e., low-flow habitats [47]), thereby having a stronger impact on coral health for a smaller proportion of the population. Finally, different coral species have important effects on the biodiversity and function of resident fishes, with several colony structure traits directly associated with fish-derived services (i.e., hosting fish, retention of nutrients [25]).

By analyzing the occupancy and biomass of damselfish, one of the most abundant reef fish families that make an important contribution to reef food webs [34], this research illustrates that both large-scale features of reef habitats and fine-scale coral morphological traits contribute to fish-coral association. Several coral-dwelling damselfish species are constrained to certain reef habitats likely due to the physical constraints of the habitat, such as high-water energy. However, even after accounting for extrinsic factors, there are important colony traits that influence colony use. Clearly, studies of coral-associated fauna across multiple spatial scales [30] that go beyond simply quantifying fauna presence–absence are necessary to understand the population dynamics of corals and symbiotic fauna. Quantifying the establishment and maintenance of such symbiotic associations with scleractinian corals will be essential to predicting how these complex networks operate under global environmental stress [83]. Indeed, many of these branching coral species, particularly, *Seriatopora hystrix*, which hosts the highest damselfish biomass, are the species most vulnerable to global climate change [51,106]; the loss of these coral species will reduce considerably the habitat for small-bodied fishes [22]. Moreover, the high degree of spatial variation in the strength of fish–coral interactions and other symbiotic interactions will make it challenging to predict their ecological functioning and cost-benefit ratios.

Supplementary Materials: The following are available online at http://www.mdpi.com/1424-2818/11/11/219/s1; Figure S1: Focal coral-fish interactions of this study focused on (a–e) five common small-branching coral species, Figure S2: Illustration of 7 coral colony structure attributes for five species of branching colonies (15–100 cm diameter) for 226 colonies over 15 sites on 11 reefs, Figure S3: Average biomass (g ± SE) of damselfishes per occupied colony on the different reef habitat zones, Figure S4: Principal component analysis (PCA) of coral colony structure attributes for n = 216 branching corals with and without resident damselfishes along mid-shelf and off-shore reefs of the GBR; Table S1: Relative importance of environmental variables influencing fish-coral interactions (reef seascape level), based on *MuMIn* model selection and model averaging, with Akaike information criteria (AICc) weighting schemes, Table S2: Binomial generalized linear model (GLM) output for fishes (species pooled) occupation by coral species (reef seascape level dataset), Table S3: Descriptive statistics of reef seascape biomass estimated (mean grams ± SE) for each damselfish species and total biomass pooled for all coral species (per occupied colony of *Acropora intermedia*, *Acropora spathulata*, *Pocillopora damicornis*, *Seriatopora hystrix*, and *Stylophora pistillata*) by site aspect (sheltered or exposed), Table S4: Descriptive statistics of reef seascape biomass estimated (mean grams ± SE) for each damselfish species (*Chromis viridis*, *Dascyllus aruanus*, *Dascyllus reticulatus*, *Pomacentrus amboinensis*, and *Pomacentrus moluccensis*) and total biomass pooled for all coral species, Table S5: Average reef seascape biomass estimates (mean ± SE) for each damselfish species (*Chromis viridis*, *Dascyllus aruanus*, *Dascyllus reticulatus*, *Pomacentrus amboinensis*, and *Pomacentrus moluccensis*) on each coral species, Table S6: Multiple comparisons of coral-species, with p-values, (Tukey's HSD post hoc) based on a Gaussian generalized linear model of total damselfish biomass, Table S7: Multiple coral species comparisons with p-values (post hoc Dunn test for (Benjamini–Hochberg method based off a Kruskal–Wallis rank sum test) for each damselfish species (damselfish-species specific biomass) for only occupied colonies, Table S8: Tukey's HSD post hoc test for multiple comparisons of position of coral on benthos, with p-values, based on a binomial generalized linear model of damselfish presence with damselfish species pooled, Table S9: Tukey's HSD post hoc test for multiple comparisons of position of coral on benthos, with p-values, based on a lognormal linear model of total biomass with damselfish species pooled for only occupied colonies, Table S10: Variance explained and linear models displaying differences between coral colonies with and without fish along principal component analyses PC1 and PC2, for a subset of coral colonies ($n = 226$) at 15 different sites on 11 reefs, Table S11: Similarity percentage analysis (SIMPER) results displaying the cumulative contributions of the most influential colony structure variables on coral colony occupation (presence or absence) by damselfishes, Table S12: Series of linear models illustrating variation in damselfishes' occupancies on small-branching coral colonies.

Author Contributions: Conceptualization, methodology, data collection, formal analysis, resources, data curation, writing—original draft preparation, writing—review and editing, visualization, project administration, funding acquisition all completed by T.JC. and M.OH.

Funding: This research was funded by the Australian Research Council to the ARCCOE for Coral Reef Studies CE140100020 and James Cook University. The funders have no role in study design, data collection and analysis, decision to publish, or preparation of the manuscript.

Acknowledgments: We thank Lizard Island staff, Grace Frank, Margaux Hein, Saskia Jurriaans, Sterling Tebbett, Andrew Baird, and the crew of the RV Kalinda for their field support and assistance. This project was implemented in accordance with the Great Barrier Reef Marine Park Authority permit (G15/37657.1, G15/37950.1, and G16/38437.1), James Cook University Animal Ethics Permit (A2186 and A2207), and James Cook University's General Fisheries Permit (170251).

Conflicts of Interest: The authors declare no conflict of interest.

References

1. Graham, N.A.J.; Nash, K.L. The importance of structural complexity in coral reef ecosystems. *Coral Reefs* **2013**, *32*, 315–326. [CrossRef]
2. Bell, J.; Galzin, R. Influence of live coral cover on coral-reef fish communities. *Mar. Ecol. Prog. Ser.* **1984**, *15*, 265–274. [CrossRef]
3. Roberts, C.; Ormond, R. Habitat complexity and coral reef fish diversity and abundance on Red Sea fringing reefs. *Mar. Ecol. Prog. Ser.* **1987**, *41*, 1–8. [CrossRef]
4. Friedlander, A.M.; Parrish, J.D. Habitat characteristics affecting fish assemblages on a Hawaiian coral reef. *J. Exp. Mar. Bio. Ecol.* **1998**, *224*, 1–30. [CrossRef]
5. Sale, P. Extremely Limited Home Range in a Coral Reef Fish, *Dascyllus aruanus* (Pisces; Pomacentridae). *Copeia* **1971**, *1971*, 324–327. [CrossRef]
6. Munday, P.L.; Jones, G.P.; Caley, M.J. Habitat specialisation and the distribution and abundance of coral-dwelling gobies. *Mar. Ecol. Prog. Ser.* **1997**, *152*, 227–239. [CrossRef]
7. Wilson, S.; Burgess, S.; Cheal, A.; Emslie, M.; Fisher, R.; Miller, I.; Polunin, N. Habitat utilization by coral reef fish: implications for specialists vs. generalists in a changing environment. *J. Anim. Ecol.* **2007**, *77*, 220–228. [CrossRef]
8. Feary, D.A. The influence of resource specialization on the response of reef fish to coral disturbance. *Mar. Biol.* **2007**, *153*, 153–161. [CrossRef]
9. Komyakova, V.; Munday, P.L.; Jones, G.P. Relative importance of coral cover, habitat complexity and diversity in determining the structure of reef fish communities. *PLoS ONE* **2013**, *8*, e83178. [CrossRef]
10. Graham, N.A.J.; Wilson, S.K.; Jennings, S.; Polunin, N.V.C.; Bijoux, J.P.; Robinson, J. Dynamic fragility of oceanic coral reef ecosystems. *Proc. Natl. Acad. Sci. USA* **2006**, *103*, 8425–8429. [CrossRef]
11. Wilson, S.K.; Graham, N.A.J.; Pratchett, M.S.; Jones, G.P.; Polunin, N.V.C. Multiple disturbances and the global degradation of coral reefs: Are reef fishes at risk or resilient? *Glob. Chang. Biol.* **2006**, *12*, 2220–2234. [CrossRef]
12. Kramer, M.J.; Bellwood, D.R.; Taylor, R.B.; Bellwood, O. Benthic crustacea from tropical and temperate reef locations: Differences in assemblages and their relationship with habitat structure. *Coral Reefs* **2017**, *36*, 971–980. [CrossRef]
13. Messmer, V.; Jones, G.P.; Munday, P.L.; Holbrook, S.J.; Schmitt, R.J.; Brooks, A.J. Habitat biodiversity as a determinant of fish community structure on coral reefs. *Ecology* **2011**, *92*, 2285–2298. [CrossRef] [PubMed]

14. Beukers, J.; Jones, J. Habitat complexity modifies the impact of piscivores on a coral reef fish population. *Oecologia* **1997**, *114*, 50–59. [CrossRef]

15. Almany, G.R. Differential effects of habitat complexity, predators and competitors on abundance of juvenile and adult coral reef fishes. *Oecologia* **2004**, *141*, 105–113. [CrossRef]

16. Munday, P.; Jones, G.; Caley, J. Interspecific competition and coexistence in a guild of coral-dwelling fishes. *Ecology* **2001**, *82*, 2177–2189. [CrossRef]

17. Cole, A.J.; Pratchett, M.S.; Jones, G.P. Diversity and functional importance of coral-feeding fishes on tropical coral reefs. *Fish Fish* **2008**, *9*, 286–307. [CrossRef]

18. Emslie, M.J.; Logan, M.; Cheal, A.J. The Distribution of planktivorous damselfishes (Pomacentridae) on the Great Barrier Reef and the relative influences of habitat and predation. *Diversity* **2019**, *11*, 33. [CrossRef]

19. Jones, G.P.; McCormick, M.I.; Srinivasan, M.; Eagle, J.V. Coral decline threatens fish biodiversity in marine reserves. *Proc. Natl. Acad. Sci. USA* **2004**, *101*, 8251–8253. [CrossRef]

20. Booth, D.J.; Wellington, G. Settlement preferences in coral-reef fishes: Effects on patterns of adult and juvenile distributions, individual fitness and population structure. *Aust. J. Ecol.* **1998**, *23*, 274–279. [CrossRef]

21. Holbrook, S.J.; Forrester, G.E.; Schmitt, R.J. Spatial patterns in abundance of damselfish reflect availability of suitable habitat. *Oecologia* **2000**, *122*, 109–120. [CrossRef] [PubMed]

22. Pratchett, M.S.; Coker, D.J.; Jones, G.P.; Munday, P.L. Specialization in habitat use by coral reef damselfishes and their susceptibility to habitat loss. *Ecol. Evol.* **2012**, *2*, 2168–2180. [CrossRef] [PubMed]

23. Munday, P.L. Does habitat availability determine geographical-scale abundances of coral-dwelling fishes? *Coral Reefs* **2002**, *21*, 105–116. [CrossRef]

24. Mellin, C.; Bradshaw, C.; Meekan, M.; Caley, M. Environmental and spatial predictors of species richness and abundance in coral reef fishes. *Glob. Ecol. Biogeogr.* **2019**, *19*, 212–222. [CrossRef]

25. Darling, E.S.; Graham, N.A.J.; Januchowski-Hartley, F.A.; Nash, K.L.; Pratchett, M.S.; Wilson, S.K. Relationships between structural complexity, coral traits, and reef fish assemblages. *Coral Reefs* **2017**, *36*, 561–575. [CrossRef]

26. Holbrook, S.J.; Brooks, A.J.; Schmitt, R.J.; Stewart, H.L. Effects of sheltering fish on growth of their host corals. *Mar. Biol.* **2008**, *155*, 521–530. [CrossRef]

27. Chase, T.; Pratchett, M.; Frank, G.; Hoogenboom, M. Coral-dwelling fish moderate bleaching susceptibility of coral hosts. *PLoS ONE* **2018**, *13*, e0208545. [CrossRef]

28. Meyer, J.; Schultz, E.; Helfman, G. Fish Schools: An Asset to Corals. *Science* **1983**, *220*, 1047–1049. [CrossRef]

29. Kane, C.N.; Brooks, A.J.; Holbrook, S.J.; Schmitt, R.J. The role of microhabitat preference and social organization in determining the spatial distribution of a coral reef fish. *Environ. Biol. Fishes* **2009**, *84*, 1–10. [CrossRef]

30. Nash, K.L.; Graham, N.A.J.; Wilson, S.K.; Bellwood, D.R. Cross-scale habitat structure drives fish body size distributions on coral reefs. *Ecosystems* **2013**, *16*, 478–490. [CrossRef]

31. Epstein, H.E.; Kingsford, M.J. Are soft coral habitats unfavourable? A closer look at the association between reef fishes and their habitat. *Environ. Biol. Fishes* **2019**, *102*, 479–497. [CrossRef]

32. Bay, L.; Jones, G.; McCormick, M. Habitat selection and aggression as determinants of spatial segregation among damselifsh on a coral reef. *Coral Reefs* **2001**, *20*, 289–298.

33. Sale, P.F. Appropriate spatial scales for studies of reef-fish ecology. *Austral. Ecol.* **1998**, *23*, 202–208. [CrossRef]

34. Frédérich, B.; Fabri, G.; Lepoint, G.; Vandewalle, P.; Parmentier, E. Trophic niches of thirteen damselfishes (Pomacentridae) at the Grand Récif of Toliara, Madagascar. *Ichthyol. Res.* **2009**, *56*, 10–17. [CrossRef]

35. Gajdzik, L.; Parmentier, E.; Michel, L.N.; Sturaro, N.; Soong, K.; Lepoint, G.; Frédérich, B. Similar levels of trophic and functional diversity within damselfish assemblages across Indo-Pacific coral reefs. *Funct. Ecol.* **2018**, *32*, 1358–1369. [CrossRef]

36. McMahon, K.W.; Thorrold, S.R.; Houghton, L.A.; Berumen, M.L. Tracing carbon flow through coral reef food webs using a compound-specific stable isotope approach. *Oecologia* **2016**, *180*, 809–821. [CrossRef]

37. Wyatt, A.S.J.; Waite, A.M.; Humphries, S. Stable isotope analysis reveals community-level variation in fish trophodynamics across a fringing coral reef. *Coral Reefs* **2012**, *31*, 1029–1044. [CrossRef]

38. Gajdzik, L.; Parmentier, E.; Sturaro, N.; Frédérich, B. Trophic specializations of damselfishes are tightly associated with reef habitats and social behaviours. *Mar. Biol.* **2016**, *163*, 1–15. [CrossRef]

39. Nadler, L.E.; McNeill, D.C.; Alwany, M.A.; Bailey, D.M. Effect of habitat characteristics on the distribution and abundance of damselfish within a Red Sea reef. *Environ. Biol. Fishes* **2014**, *97*, 1265–1277. [CrossRef]

40. Fulton, C.J.; Bellwood, D.R.; Wainwright, P.C. Wave energy and swimming performance shape coral reef fish assemblages. *Proc. R. Soc. B Biol. Sci.* **2005**, *272*, 827–832. [CrossRef]

41. Johansen, J.L.; Bellwood, D.R.; Fulton, C.J. Coral reef fishes exploit flow refuges in high-flow habitats. *Mar. Ecol. Prog. Ser.* **2008**, *360*, 219–226. [CrossRef]

42. Fulton, C.J.; Bellwood, D.R. Wave-induced water motion and the functional implications for coral reef fish assemblages. *Limnol. Oceanogr.* **2005**, *50*, 255–264. [CrossRef]

43. Sambrook, K.; Jones, G.P.; Bonin, M.C. Life on the edge: Coral reef fishes exhibit strong responses to a habitat boundary. *Mar. Ecol. Prog. Ser.* **2016**, *561*, 203–215. [CrossRef]

44. Kerry, J.T.; Bellwood, D.R. The effect of coral morphology on shelter selection by coral reef fishes. *Coral Reefs* **2012**, *31*, 415–424. [CrossRef]

45. Pratchett, M.S.; Baird, A.H.; McCowan, D.M.; Coker, D.J.; Cole, A.J.; Wilson, S.K. Protracted declines in coral cover and fish abundance following climate-induced coral bleaching on the Great Barrier Reef. In Proceedings of the 11th International Coral Reef Symposium, Fort Lauderdale, FL, USA, 7–11 July 2008; pp. 1309–1313.

46. Coker, D.; Wilson, S.; Pratchett, M. Importance of live coral habitat for reef fishes. *Rev. Fish Biol. Fish.* **2014**, *24*, 89–126. [CrossRef]

47. Chase, T.; Pratchett, M.; Walker, S.; Hoogenboom, M. Small-scale environmental variation influences whether coral-dwelling fish promote or impede coral growth. *Oecologia* **2014**, *176*, 1009–1022. [CrossRef]

48. Randall, J.; Allen, G.; Steene, R. *Fishes of the Great Barrier Reef and Coral Sea*; University of Hawai'i Press: Honolulu, HI, USA, 1997.

49. Allen, G.; Steene, R.; Humann, P.; DeLoach, N. *Reef Fish Identification—Tropical Pacific*; New World Pulications, Inc.: Jacksonville, FL, USA, 2003.

50. Chase, T.; Pratchett, M.; Hoogenboom, M. Behavioural trade-offs and habitat associations of coral-dwelling damselfishes (family Pomacentridae). *Mar. Ecol. Prog. Ser.* **2019**. [CrossRef]

51. Hughes, T.P.; Kerry, J.T.; Álvarez-Noriega, M.; Álvarez-Romero, J.G.; Anderson, K.D.; Baird, A.H.; Babcock, R.C.; Beger, M.; Bellwood, D.R.; Berkelmans, R.; et al. Global warming and recurrent mass bleaching of corals. *Nature* **2017**, *543*, 373–377. [CrossRef]

52. Hoogenboom, M.O.; Frank, G.E.; Chase, T.J.; Jurriaans, S.; Álvarez-Noriega, M.; Peterson, K.; Critchell, K.; Berry, K.L.E.; Nicolet, K.J.; Ramsby, B.; et al. Environmental drivers of variation in bleaching severity of Acropora species during an extreme thermal anomaly. *Front. Mar. Sci.* **2017**, *4*, 376. [CrossRef]

53. Graham, N.A.J.; Chong-Seng, K.M.; Huchery, C.; Januchowski-Hartley, F.A.; Nash, K.L. Coral reef community composition in the context of disturbance history on the Great Barrier Reef, Australia. *PLoS ONE* **2014**, *9*. [CrossRef]

54. Hopley, D.; Smithers, S.; Parnell, K. *The Geomorphology of the Great Barrier Reef: Development, Diversity and Change*; Cambridge University Press: Cambridge, UK, 2007.

55. Ménard, A.; Turgeon, K.; Roche, D.G.; Binning, S.A.; Kramer, D.L. Shelters and their use by fishes on fringing coral reefs. *PLoS ONE* **2012**, *7*. [CrossRef]

56. Veron, J.; Stafford-Smith, M. *Corals of the World*; Australian Institute of Marine Science: Townsville, Australia, 2004.

57. Hill, J.; Wilkinson, C. *Methods for Ecological Monitoring of Coral Reefs*; Australian Institute of Marine Science: Townsville, Australia, 2004; 117p.

58. Wehrberger, F.; Herler, J. Microhabitat characteristics influence shape and size of coral-associated fishes. *Mar. Ecol. Prog. Ser.* **2014**, *500*, 203–214. [CrossRef]

59. Fernandes, I.M.; Bastos, Y.F.; Barreto, D.S.; Lourenço, L.S.; Penha, J.M. The efficacy of clove oil as an anaesthetic and in euthanasia procedure for small-sized tropical fishes. *Braz. J. Biol.* **2016**, *77*, 444–450. [CrossRef]

60. Javahery, S.; Nekoubin, H.; Moradlu, A.H. Effect of anaesthesia with clove oil in fish (review). *Fish Physiol. Biochem.* **2012**, *38*, 1545–1552. [CrossRef]

61. Burkepile, D.E.; Allgeier, J.E.; Shantz, A.A.; Pritchard, C.E.; Lemoine, N.P.; Bhatti, L.H.; Layman, C.A. Nutrient supply from fishes facilitates macroalgae and suppresses corals in a Caribbean coral reef ecosystem. *Sci. Rep.* **2014**, *3*, 19–21. [CrossRef]

62. Cribari-Neto, F.; Zeileis, A. Journal of Statistical Software Beta Regression in R. *J. Stat. Softw.* **2010**, *34*, 1–24. [CrossRef]

63. Warton, D.I.; Hui, F. The arcsine is asinine: The analysis of proportions in ecology. *Ecology* **2011**, *92*, 3–10. [CrossRef]
64. Affleck, D.L.R. Additivity and maximum likelihood estimation of nonlinear component biomass models. In Proceedings of the New Directions in Inventory Techinques & Applicaitons Forest Inventory & Analysis (FIA) Symposium 2015, 8–12 December 2015; U.S. Department of Agriculture, Forest Service, Pacific Northwest Research Station: Portland, OR, USA; pp. 13–17.
65. Zuur, A.; Ieno, E.; Walker, N.; Saveliev, A.; Smith, G. *Mixed Effects Models and Extension in Ecology with R*; Springer: New York, NY, USA, 2009.
66. Zuur, A.F.; Ieno, E.N. A protocol for conducting and presenting results of regression-type analyses. *Methods Ecol. Evol.* **2016**, *7*, 636–645. [CrossRef]
67. Bartón, K. *MuMIN: Multi-model Interference R Package Version 1.9.12*; The R Foundation: Vienna, Austria, 2013.
68. Burnham, K.P.; Anderson, D.R. *Multimodel Inference: Understanding AIC and BIC in Model Selection*, 2nd ed.; Springer: New York, NY, USA, 2002.
69. Benjamini, Y.; Hochber, Y. Controlling the false discovery rate: A practical and powerful approach to multiple testing. *J. R. Stat. Soc. B* **1995**, *57*, 289–300. [CrossRef]
70. Venables, W.; Ripley, B. *Modern Applied Statistics with S*, 4th ed.; Springer: Berlin, Germany, 2002.
71. Mardia, K.; Kent, J.; Bibby, J. *Multivariate Analysis*; Academic Press: London, UK, 1979.
72. Clarke, K.R. Non-parametric multivariate analyses of changes in community structure. *Aust. J. Ecol.* **1993**, *18*, 117–143. [CrossRef]
73. Warton, D.I.; Wright, S.T.; Wang, Y. Distance-based multivariate analyses confound location and dispersion effects. *Methods Ecol. Evol.* **2012**, *3*, 89–101. [CrossRef]
74. R Development Core Team. *R: A Language and Environment for Statistical Computing*; The R Foundation: Vienna, Austria, 2018.
75. Hothorn, T.; Bretz, F.; Westfall, P. Simulatnous Inference in Generall Parametric Models. *Biom. J.* **2008**, *50*, 346–363. [CrossRef] [PubMed]
76. Lenth, R.V. Least-Squares Means: The R Package lsmeans. *J. Stat. Softw.* **2016**, *69*, 1–33. [CrossRef]
77. Chase, T.; Hoogenboom, M. Occupation of Damselfishes across Reef Seascape and Colony Scale, GBR 2016 Data. James Cook University (Dataset). Available online: http://dx.doi.org/10.25903/5dcb4c44aa86a (accessed on 19 November 2019).
78. Noonan, S.H.C.; Jones, G.P.; Pratchett, M.S. Coral size, health and structural complexity: Effects on the ecology of a coral reef damselfish. *Mar. Ecol. Prog. Ser.* **2012**, *456*, 127–137. [CrossRef]
79. Pereira, P.H.C.; Munday, P.L. Coral colony size and structure as determinants of habitat use and fitness of coral-dwelling fishes. *Mar. Ecol. Prog. Ser.* **2016**, *553*, 163–172. [CrossRef]
80. Waldner, R.E.; Robertson, D.R. Patterns of habitat partitioning by eight species of territorial caribbean damselfishes (Pisces: Pomacentridae). *Bull. Mar. Sci.* **1980**, *30*, 171–186.
81. Morais, R.A.; Bellwood, D.R. Pelagic Subsidies Underpin Fish Productivity on a Degraded Coral Reef. *Curr. Biol.* **2019**, *29*, 1521–1527.e6. [CrossRef]
82. Meekan, M.G.; Steven, A.D.L.; Fortin, M.J. Spatial patterns in the distribution of damselfishes on a fringing coral reef. *Coral Reefs* **1995**, *14*, 151–161. [CrossRef]
83. Waldock, C.; Stuart-Smith, R.D.; Edgar, G.J.; Bird, T.J.; Bates, A.E. The shape of abundance distributions across temperature gradients in reef fishes. *Ecol. Lett.* **2019**, *22*, 685–696. [CrossRef]
84. Ault, T.; Johnson, C. Spatial variation in fish species richness on coral reefs: Habitat fragmentation and stochastic structuring processes. *Okios* **1998**, *82*, 354–364. [CrossRef]
85. Bergman, K.C.; Öhman, M.C.; Svensson, S. Influence of habitat structure on *Pomacentrus sulfureus*, a western Indian Ocean reef fish. *Environ. Biol. Fishes* **2000**, *59*, 243–252. [CrossRef]
86. Maire, E.; Villeger, S.; Graham, N.A.J.; Hoey, A.S.; Cinner, J.; Ferse, S.C.A.; Aliaume, C.; Booth, D.J.; Feary, D.A.; Kulbicki, M.; et al. Community-wide scan identifies fish species associated with coral reef services across the Indo-Pacific. *Proc. R. Soc. B Biol. Sci.* **2018**, *285*, 1621–1629. [CrossRef]
87. Doherty, P.; Fowler, T. An empirical test of recruitment limitation in a coral reef fish. *Science* **1994**, *263*, 935–939. [CrossRef]
88. Forrester, G. Strong density-dependent survival and recruitment regulate the abundance of a coral reef fish. *Oecologia* **1995**, *103*, 275–282. [CrossRef]

89. Holbrook, S.J.; Schmitt, R.J. Spatial and temporal variation in mortality of newly settled damselfish: Patterns, causes and co-variation with settlement. *Oecologia* **2003**, *135*, 532–541. [CrossRef]

90. Motro, R.; Ayalon, I.; Genin, A. Near-bottom depletion of zooplankton over coral reefs: III: Vertical gradient of predation pressure. *Coral Reefs* **2005**, *24*, 95–98. [CrossRef]

91. Zikova, A.V.; Britaev, T.A.; Ivanenko, V.N.; Mikheev, V.N. Planktonic and symbiotic organisms in nutrition of coralobiont fish. *J. Ichthyol.* **2011**, *51*, 769–775. [CrossRef]

92. Chamberlain, S.A.; Bronstein, J.L.; Rudgers, J.A. How context dependent are species interactions? *Ecol. Lett.* **2014**, *17*, 881–890. [CrossRef]

93. Nanami, A.; Nishihira, M. Effects of habitat connectivity on the abundance and species richness of coral reef fishes: Comparison of an experimental habitat established at a rocky reef flat and at a sandy sea bottom. *Environ. Biol. Fishes* **2003**, *68*, 183–196. [CrossRef]

94. Shpigel, M.; Fishelson, L. Behavior and physiology of coexistence in two species of *Dascyllus* (Pomacentridae, Teleostei). *Environ. Biol. Fishes* **1986**, *17*, 253–265. [CrossRef]

95. Sale, P.F. Effect of cover on agonistic behavior of a reef fish: A possible spacing mechanism. *Ecolgoy* **1972**, *53*, 753–758. [CrossRef]

96. Sale, P.F. Influence of corals in the dispersion of the Pomacentrid fish, *Dascyllus aruanus*. *Ecology* **1972**, *53*, 741–744. [CrossRef]

97. Pratchett, M.; Hoey, A.; Wilson, S.; Hobbs, J.; Allen, G. Habitat-use and specialisation among coral reef damselfishes. In *Biology of Damselfishes*; Frédérich, B., Parmentier, E., Eds.; CRC Press: Boca Raton, FL, USA, 2016; pp. 84–121.

98. Eurich, J.; McCormick, M.; Jones, G. Habitat selection and aggression as determinants of fine-scale partitioning of coral reef zones in a guild of territorial damselfishes. *Mar. Ecol. Prog. Ser.* **2018**, *587*, 201–215. [CrossRef]

99. Almany, G.R. Does increased habitat complexity reduce predation and competition in coral reef fish assemblages? *Oikos* **2004**, *106*, 275–284. [CrossRef]

100. Wen, C.K.C.; Pratchett, M.S.; Almany, G.R.; Jones, G.P. Patterns of recruitment and microhabitat associations for three predatory coral reef fishes on the southern Great Barrier Reef, Australia. *Coral Reefs* **2013**, *32*, 389–398. [CrossRef]

101. Chase, T. *Effects of Coral-Dwelling Damselfishes' Abundances and Diversity on Host Coral Dynamics*; James Cook University: Townsvile, Australia, 2019.

102. Shantz, A.A.; Ladd, M.C.; Schrack, E.; Burkepile, D.E. Fish-derived nutrient hotspots shape coral reef benthic communities. *Ecol. Appl.* **2015**, *25*, 2142–2152. [CrossRef]

103. Garcia-Herrera, N.; Ferse, S.C.A.; Kunzmann, A.; Genin, A. Mutualistic damselfish induce higher photosynthetic rates in their host coral. *J. Exp. Biol.* **2017**, *220*, 1803–1811. [CrossRef]

104. Meyer, J.L.; Schultz, E.T. Migrating haemulid fishes as a source of nutrients and organic matter on coral reefs. *Limnol. Oceanogr.* **1985**, *30*, 146–156. [CrossRef]

105. Meyer, J.; Schultz, E. Tissue condition and growth rate of corals associated with schooling fish. *Limnol. Oceanogr.* **2019**, *30*, 157–166. [CrossRef]

106. Hoegh-Guldberg, O.; Smith, G.J. The effect of sudden changes in temperature, light and salinity on the population density and export of zooxanthellae from the reef corals *Stylophora pistillata* Exper and *Seriatopora hystrix* Dana. *J. Exp. Mar. Biol. Ecol.* **1989**, *129*, 279–303. [CrossRef]

 *diversity*

Article

Green Fluorescence Patterns in Closely Related Symbiotic Species of *Zanclea* (Hydrozoa, Capitata)

Davide Maggioni [1,2,*], Luca Saponari [1,2], Davide Seveso [1,2], Paolo Galli [1,2], Andrea Schiavo [1,2], Andrew N. Ostrovsky [3,4] and Simone Montano [1,2]

[1]   Department of Earth and Environmental Sciences (DISAT), University of Milano-Bicocca, 20126 Milano, Italy; luca.saponari@unimib.it (L.S.); davide.seveso@unimib.it (D.S.); paolo.galli@unimib.it (P.G.); a.schiavo@campus.unimib.it (A.S.); simone.montano@unimib.it (S.M.)
[2]   Marine Research and High Education (MaRHE) Center, University of Milano-Bicocca, Faafu Magoodhoo 12030, Maldives
[3]   Department of Palaeontology, University of Vienna, 1090 Vienna, Austria; a.ostrovsky@spbu.ru
[4]   Department of Invertebrate Zoology, Saint Petersburg State University, 199034 Saint Petersburg, Russia
*   Correspondence: davide.maggioni@unimib.it

Received: 7 January 2020; Accepted: 17 February 2020; Published: 18 February 2020

Abstract: Green fluorescence is a common phenomenon in marine invertebrates and is caused by green fluorescent proteins. Many hydrozoan species display fluorescence in their polyps and/or medusa stages, and in a few cases patterns of green fluorescence have been demonstrated to differ between closely related species. Hydrozoans are often characterized by the presence of cryptic species, due to the paucity of available morphological diagnostic characters. *Zanclea* species are not an exception, showing high genetic divergence compared to a uniform morphology. In this work, the presence of green fluorescence and the morpho-molecular diversity of six coral- and bryozoan-associated *Zanclea* species from the Maldivian coral reefs were investigated. Specifically, the presence of green fluorescence in polyps and newly released medusae was explored, the general morphology, as well as the cnidome and the interaction with the hosts, were characterized, and the *16S rRNA* region was sequenced and analyzed. Overall, *Zanclea* species showed a similar morphology, with little differences in the general morphological features and in the cnidome. Three of the analyzed species did not show any fluorescence in both life stages. Three other *Zanclea* species, including two coral-associated cryptic species, were distinguished by species-specific fluorescence patterns in the medusae. Altogether, the results confirmed the morphological similarity despite high genetic divergence in *Zanclea* species and indicated that fluorescence patterns may be a promising tool in further discriminating closely related and cryptic species. Therefore, the assessment of fluorescence at a large scale in the whole Zancleidae family may be useful to shed light on the diversity of this enigmatic taxon.

Keywords: integrative taxonomy; symbiosis; corals; bryozoans; Maldives; phylogeny

---

1. Introduction

Green fluorescence is a diffuse phenomenon in the marine environment, being found in a variety of taxa, including cnidarians, ctenophores, crustaceans, and chordates [1]. Green fluorescence is caused by green fluorescence proteins, which were firstly described in the hydrozoan species *Aequorea victoria* (Murbach and Shearer, 1902) [2]. Lately, similar proteins were detected in several other species, mainly belonging to the Anthozoa [3], and they are currently known to be widespread in the marine metazoans [4]. In most cases, the ecological function of fluorescence is still unclear, even though some hypotheses have been proposed. For instance, in anthozoans associated with unicellular algae, fluorescent proteins may have a role in regulating the light environment of the symbionts [5], whereas in bioluminescent organisms they seem to be involved in the modification of bioluminescence emission [6]. However, these hypotheses do not apply to non-symbiotic (with algae) and non-bioluminescent species. Other possible roles of fluorescence in marine organisms relate to camouflage, intraspecific communication [7], and prey attraction [8,9]. The latter hypothesis seems to fit better for hydrozoans, since it has been experimentally demonstrated that at least one species, *Olindias formosus* (Goto, 1903), uses fluorescence in tentacles to attract juvenile fish preys [8].

Among hydrozoans, green fluorescence is common and has been reported from polyps and medusae of several species (see [10] and references therein). In medusae, fluorescence is found in the umbrella, radial and circular canals, manubrium, gonads, bulbs, and tentacles (e.g., [11–13]), whereas in polyps in the hydrocaulus, hypostome, and in the epithelium below tentacles [10,13,14]. Green fluorescence patterns were found to differ significantly in closely related species of *Eugymnanthea* Palombi, 1936 [11], and even if these patterns changed during the development, they remained distinguishable from those in the relatives [11]. Moreover, Prudkovsky et al. [10] recently demonstrated that these patterns also differ between cryptic or pseudo-cryptic species of *Cytaeis* Eschscholtz, 1829, indicating that they may be reliable and informative taxonomic characters that could be useful especially when dealing with morphologically undistinguishable species.

Indeed, cryptic species are common in hydrozoans, since morphologically very similar polyps and medusae often show strong genetic diversification, that in many cases relates to host specialization and geography (e.g., [15–17]). This is especially true for the capitate family Zancleidae Russel, 1953, in which the few morphological diagnostic characters available make species identification and description challenging [18,19]. The cnidome is considered a useful character to discriminate among zancleid species, due to the variation of type and size of nematocysts in different species [20]. For instance, the statistical treatment of nematocysts measurements of three *Zanclea* cryptic species resulted in significant differences between the taxa [21], further supporting the importance of the cnidome as a reliable taxonomic character. Another useful character to distinguish closely related symbiotic species is the host specificity, since some species or lineages are specifically associated with one or a few invertebrate taxa (e.g., scleractinian corals) [16,18]. Moreover, some coral-associated *Zanclea* species were found to induce modifications of the host skeletons that could be taxonomically informative [21].

In this work we analyzed the morphology (polyps, newly released medusae, and modifications of the hosts) and genetic diversity (*16S rRNA*) of six symbiotic *Zanclea* species collected in the Maldives. Yet, along with the morpho-molecular analyses, we investigated the informativeness of green fluorescence patterns of polyps and medusae to discriminate between closely related taxa.

2. Materials and Methods

2.1. Morphological Analyses and Fluorescence Essay

Colonies of symbiotic *Zanclea* species were collected in reefs around Magoodhoo Island, Faafu Atoll, Republic of the Maldives (3.0782° N, 72.9613° E), during February 2017. Six *Zanclea* species were collected: *Zanclea sango* Hirose and Hirose, 2011 and *Zanclea* sp. (Clade I, *sensu* [18]) associated with the scleractinians *Pavona varians* (Verril, 1864) and *Goniastrea* sp., respectively; *Zanclea divergens* (Boero, Bouillon, and Gravili, 2000), *Zanclea* sp. 1 and *Zanclea* sp. 2 (*sensu* [22]) associated with the bryozoans *Celleporaria vermiformis* (Waters, 1909), *Celleporaria pigmentaria* (Waters, 1909), and *Celleporaria* sp., respectively; *Zanclea* cf. *protecta* associated with the bryozoans *Parasmittina* cf. *spondylicola* and *Schizoporella* sp. For comparison, *Asyncoryne ryniensis* Warren, 1908 was included in the analyses, since it is closely related to the family Zancleidae [22]. For each *Zanclea* species three colonies were collected, whereas two colonies of *A. ryniensis* were analyzed, for a total of 20 samples. Hydrozoan colonies were collected together with their hosts using hammer and chisel, by snorkeling or SCUBA diving. Colonies were immediately transferred in bowls with seawater after diving, and they were kept in the laboratories of the Marine Research and High Education (MaRHE) Center in Magoodhoo. One colony per species had medusa buds at the time of sampling, and these colonies were reared until medusae were released. Seawater was replaced daily, approximately two hours after a feeding session with *Artemia* nauplii. Newly released medusae were reared for a few days and then anesthetized with menthol crystals and fixed with 10% formalin for further morphological analyses. Hydrozoan polyps were detached from their hosts using precision forceps and micropipettes, and they were fixed in 10% formalin and 99% ethanol for morphological and genetic analyses, respectively. Formalin-preserved polyps and medusae were analyzed using a Zeiss Axioskop 40 compound microscope to observe their general morphology and characterize their cnidome. Measurements were taken using the software ImageJ 1.52p. All pictures were taken using Canon G7X Mark II camera.

To investigate possible modifications related to the associations with hydroids, the skeletons of the hosts were analyzed under a scanning electron microscope. Specifically, fragments of the *Zanclea*-bearing bryozoan and scleractinian colonies were immersed in a 10% sodium hypochlorite solution for 6–24 h. After rinsing, fragments were sputter-coated with gold and observed under a Zeiss Gemini SEM 500 scanning electron microscope.

Before fixation, all hydrozoan polyps (n = 15 for each species and colony) and medusae (n = 5–15 for each species) were checked for green fluorescence emission using a Leica EZ4 D stereomicroscope equipped with a Weefine Smart Focus 2300 lamp (excitation wavelength: 420 nm) and yellow filter. All medusae were observed at day one and five after release.

2.2. Molecular Characterization

Genetic analyses were performed to check the molecular identity of the samples (n = 20) and to assess their phylogenetic relationships. DNA was extracted from one polyp per colony using a protocol modified from Zietara et al. [23] and already used proficiently to extract DNA from hydrozoans (e.g., [24]). A portion of the *16S rRNA* was then amplified using the primers and protocol described in Cunningham and Buss [25]. The success of PCRs was assessed through an electrophoretic run in 1% agarose gel. PCR products were purified and sequenced in forward and reverse directions with the same primers used for amplification, with ABI 3730xl DNA Analyzer (Applied Biosystems). The obtained chromatograms were visually checked and assembled using Geneious 6.1.6 and sequences were deposited with the EMBL (GenBank accession numbers: MN923260-MN923279). Each sequence was searched in the NCBI BLASTn database to confirm the morphological identifications. All the obtained sequences were then aligned using MAFFT 7.110 [26], with the *E-INS-i* option and the sequences of *Cladocoryne haddoni* and *Pennaria disticha* (GenBank accession numbers: MG811591 and LT746002, respectively) were included as outgroups. The best-fitting evolutionary model was determined using JModelTest 2 [27] and resulted in GTR+I+G, following the Akaike Information Criterion. Phylogenetic trees were built using both Bayesian inference and maximum likelihood approaches. For Bayesian analyses, MrBayes 3.2.6 [28] was used, and four parallel Markov Chain Monte Carlo runs (MCMC) were run for 10^7 generations, trees were sampled every 1000th generation, and burn-in was set to 25%. Maximum likelihood trees were built with RAxML 8.2.9 [29] using 1000 bootstrap replicates.

Pairwise genetic distances between and within species were calculated as % uncorrected p-distances with 1000 bootstrap replicates using MEGA X [30].

3. Results

3.1. General Morphology of Polyps and Medusae

All the analyzed *Zanclea* species showed a similar morphology in both polyp and medusa stages (Figures 1–6). All polyps were colonial, cylindrical, or claviform, with a whorl of oral capitate tentacles and aboral tentacles scattered on the hydranth body wall. Bryozoan-associated species (*Zanclea divergens*, *Zanclea* cf. *protecta*, *Zanclea* sp. 1, and *Zanclea* sp. 2) were monomorphic and deprived of perisarc, whereas the scleractinian-associated *Zanclea sango* and *Zanclea* sp. (Clade I) showed polymorphic polyps, having both gastrozooids and dactylozooids, and the hydrorhiza was surrounded by a thin layer of chitinous perisarc. All species had stenotele capsules in their capitula, and apart from *Zanclea* cf. *protecta*, all had euryteles in their polyps and/or hydrorhiza. Medusa buds arose directly from the hydrorhiza in *Zanclea divergens*, *Zanclea* sp. 1, and *Zanclea* sp. 2, whereas they were borne on both gastrozooids and hydrorhiza in *Zanclea* cf. *protecta*, *Zanclea sango*, and *Zanclea* sp. (Clade I). Medusae had a bell-shaped or globular umbrella, with nematocysts scattered over the surface in all species apart from scleractinian-associated species. *Zanclea* sp. 1 and *Zanclea* sp. 2 did not have canals and exumbrellar nematocyst pouches at release, whereas all other species had four radial and one circular canal and four nematocyst pouches containing stenoteles and euryteles (the latter only in coral-associated species). Manubria were cylindrical and had stenoteles around the mouth in *Z. divergens*, *Z.* cf. *protecta*, *Zanclea sango*, and *Zanclea* sp. (Clade I). *Zanclea* sp. 1 and *Zanclea* sp. 2 had no nematocysts on the manubrium but four short oral arms. All medusae had two opposite tentacles, bearing a variable number of rounded or elongated cnidophores containing bean-shaped macrobasic euryteles.

Asyncoryne ryniensis (Figure 7) polyps had a distinct morphology, being characterized by a whorl of capitate oral tentacles and moniliform tentacles scattered on the hydranth body wall. Polyps were monomorphic and had both stenoteles and euryteles. Medusa buds were borne on the distal half of polyps. The medusa stage was very similar to that of *Zanclea* species, showing a bell-shaped umbrella, one circular and four radial canals, four exumbrellar nematocyst pouches, four bulbs, and two opposite tentacles bearing cnidophores with macrobasic euryteles inside.

Detailed characterizations of morphology and cnidome of polyps and medusae of all species are summarized in Tables 1 and 2.

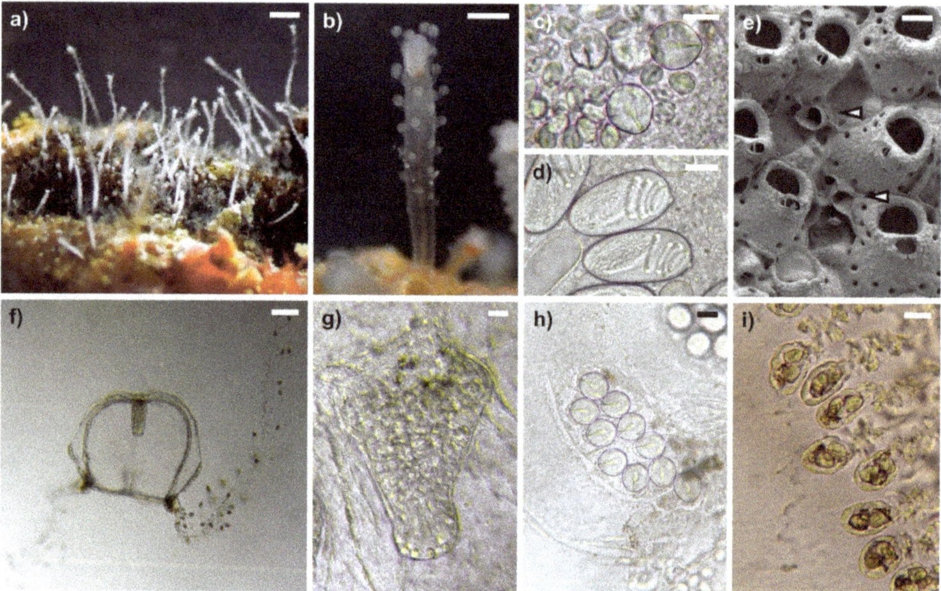

Figure 1. *Zanclea divergens*. (**a**) Colony associated with *Celleporaria vermiformis*; (**b**) close-up of a polyp; (**c**) stenoteles in the capitula, and (**d**) euryteles in the hypostome; (**e**) tube-like skeletal modifications of the bryozoan skeleton (arrowheads); (**f**) newly released medusa and close-up of (**g**) manubrium, (**h**) nematocyst pouch, and (**i**) cnidophores. Scale bars: (**a**) 0.5 mm; (**b**,**e**,**f**) 0.1 mm; (**c**,**d**,**g**–**i**) 10 μm.

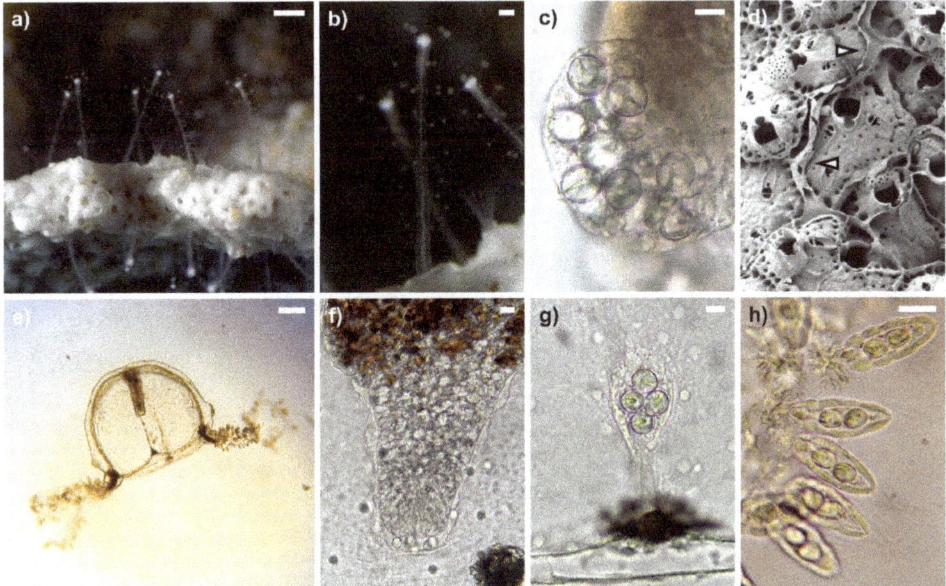

Figure 2. *Zanclea* cf. *protecta*. (**a**) Colony associated with *Parasmittina* cf. *spondylicola*; (**b**) close-up of a polyp; (**c**) stenoteles in the capitula; (**d**) bryozoan skeletal lamina overgrowing the hydrorhiza (arrowheads); (**e**) newly released medusa; close-ups of (**f**) manubrium, (**g**) nematocyst pouch, and (**h**) cnidophores. Scale bars: (**a**) 0.5 mm; (**b,d,e**) 0,1 mm; (**c,f–h**) 10 μm.

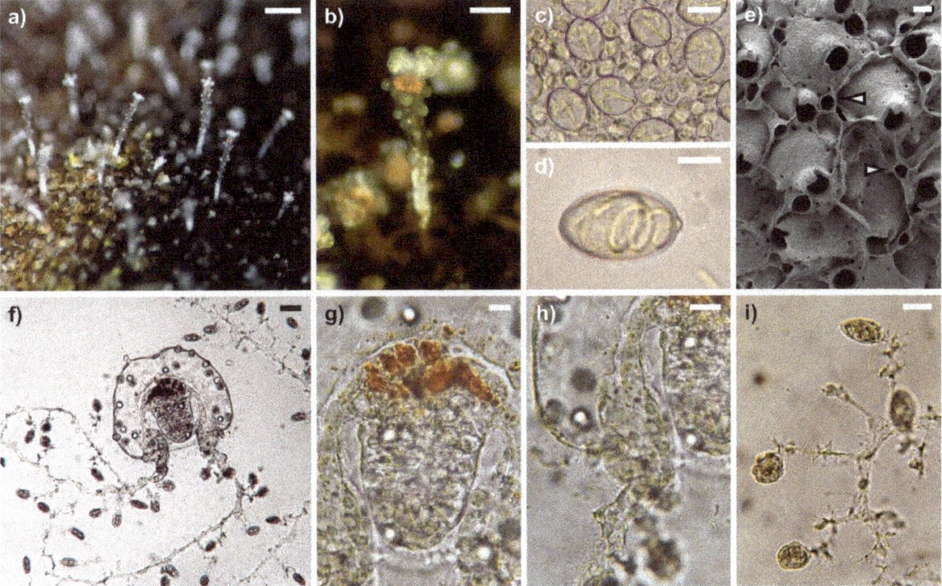

Figure 3. *Zanclea* sp. 1. (**a**) Colony associated with *Celleporaria pigmentaria*; (**b**) close-up of a polyp; (**c**) stenoteles in the capitula, and (**d**) eurytele in the hydrorhiza; (**e**) tube-like modifications of the bryozoan skeleton (arrowheads); (**f**) newly released medusa; close-ups of (**g**) manubrium, (**h**) tentacular bulb, and (**i**) cnidophores. Scale bars: (**a**) 0.5 mm; (**b,e**) 0.1 mm; (**c,d,g,h**) 10 μm; (**f**) 20 μm.

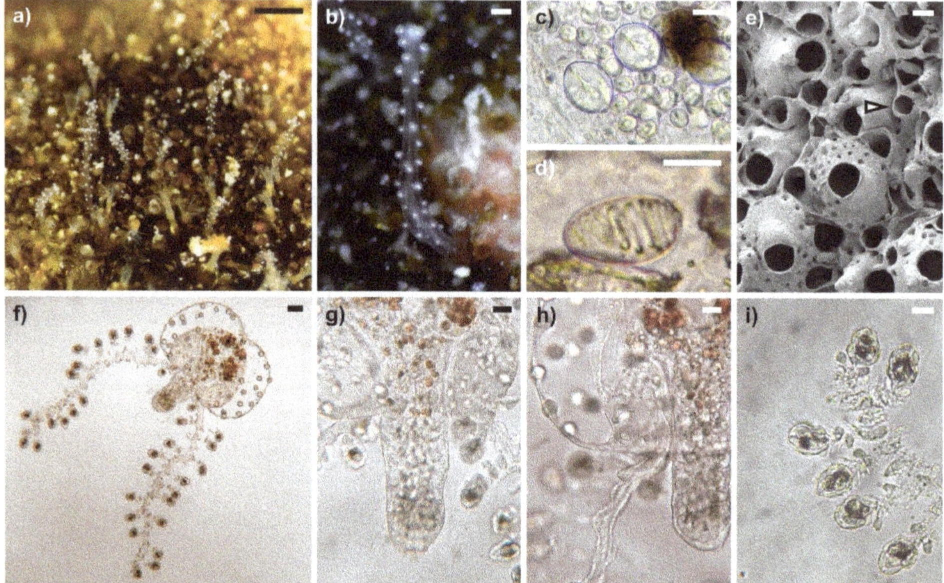

Figure 4. *Zanclea* sp. 2. (**a**) Colony associated with *Celleporaria* sp.; (**b**) close-up of a polyp; (**c**) stenoteles in the capitula, and (**d**) euryteles in the hydrorhiza; (**e**) tube-like modifications of the bryozoan skeleton (arrowheads); (**f**) newly released medusa; close-ups of (**g**) manubrium, (**h**) tentacular bulb, and (**i**) cnidophores. Scale bars: (**a**) 0.5 mm; (**b,e**) 0.1 mm; (**c,d,g–i**) 10 μm; (**f**) 20 μm.

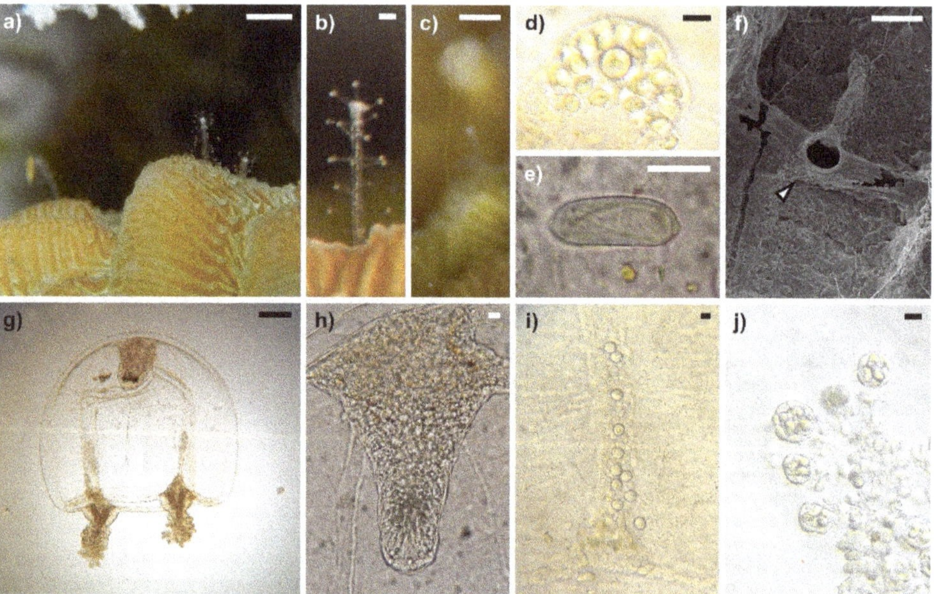

Figure 5. *Zanclea sango.* (**a**) Colony associated with *Pavona varians*; close-ups of (**b**) gastrozooid, and (**c**) dactylozooid; (**d**) stenoteles in the capitula, and (**e**) eurytele in the hypostome; (**f**) micro-alteration of the coral skeleton (arrowhead); (**g**) newly released medusa; close-ups of (**h**) manubrium, (**i**) nematocyst pouch, and (**j**) cnidophores. Scale bars: (**a,c**) 0.5 mm; (**b,f,g**) 0.1 mm; (**d,e,h–j**) 10 μm.

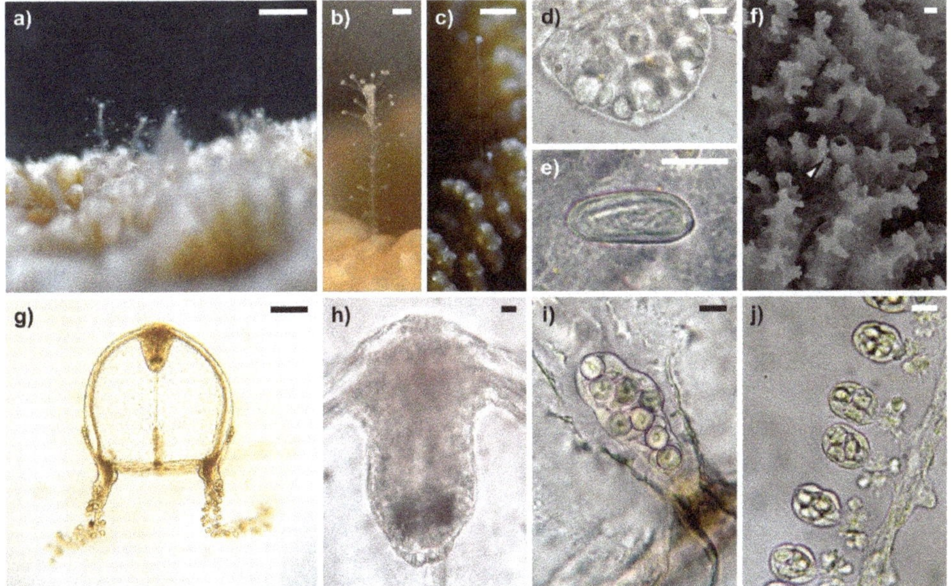

Figure 6. *Zanclea* sp. (Clade I). (**a**) Colony associated with *Goniastrea* sp.; close-ups of (**b**) gastrozooid and (**c**) dactylozooid; (**d**) stenoteles in the capitula, and (**e**) eurytele in the hypostome; (**f**) micro-alteration of the coral skeleton (arrowhead); (**g**) newly released medusa; close-ups of (**h**) manubrium, (**i**) nematocyst pouch, and (**j**) cnidophores. Scale bars: (**a**,**c**) 0.5 mm; (**b**,**f**,**g**) 0.1 mm; (**d**,**e**,**h**–**j**) 10 µm.

Figure 7. *Asyncoryne ryniensis.* (**a**) Colony growing on dead coral; (**b**) close-up of a polyp; (**c**) polyp showing green fluorescence before stimulation with blue light; (**d**) stenoteles in the capitulum, and (**e**) eurytele in the hydranth; (**f**,**g**) newly released medusa; close-ups of (**h**) manubrium, and (**i**) nematocysts in the tentacular bulb. Scale bars: (**a**) 0.1 mm; (**b**,**c**,**f**,**g**) 0.1 mm; (**d**,**e**,**h**,**i**) 10 µm.

Table 1. Morphology of the polyp stages.

Species	*Zanclea divergens*	*Zanclea* cf. *protecta*	*Zanclea* sp. 1	*Zanclea* sp. 2	*Zanclea sango*	*Zanclea* sp. (Clade I)	*Asyncoryne ryniensis*
Host/Substrate	*Celleporaria vermiformis*	*Parasmittina* cf. *spondylicola*, *Schizoporella* sp.	*Celleporaria pigmentaria*	*Celleporaria* sp.	*Pavona varians*	*Goniastrea* sp.	Rock, sponge
Hydrorhiza	Below the bryozoan skeleton, coming out in irregular notches	On the bryozoan, overgrown by the host skeleton	Below the bryozoan skeleton, coming out in irregular notches	Below the bryozoan skeleton, coming out for some distance	Below the coral skeleton and tissues	Below the coral skeleton and tissues	On the substrate
Perisarc	No	No	No	No	Yes	Yes	Yes
Gastrozooid	Cylindrical, up to 3.5 mm long	Cylindrical, up to 1.5 mm long	Claviform, up to 1.5 mm long	Cylindrical, up to 3 mm long	Cylindrical to claviform, up to 1 mm long	Cylindrical to claviform, up to 1 mm long	Cylindrical, up to 6 mm long
Gastrozooid tentacles	Capitate: 4–5 oral, 16–39 aboral	Capitate: 4–5 oral, 15–44 aboral	Capitate: 4–5 oral, 18–20 aboral	Capitate: 4–5 oral, 23–27 aboral	Capitate: 4–6 oral, 12–22 aboral	Capitate: 4–5 oral, 23–32 aboral	Capitate:3–4 oral; moniliform: 28–36 aboral
Dactylozooid	No	No	No	No	Up to 3 mm, globular apex, no tentacles	Up to 3 mm, globular apex, no tentacles	No
Medusa buds localization	Hydrorhiza	Hydrorhiza and polyps	Hydrorhiza	Hydrorhiza	Hydrorhiza and polyps	Hydrorhiza and polyps	Polyps
Color	Whitish-transparent, white hypostome	Transparent, white hypostome	Transparent, white band, whitish to orange hypostome	Transparent	Transparent, white hypostome	Transparent, white hypostome	Transparent, orange gastroderm, white hypostome
Stenoteles	Capitula	Capitula, hydrorhiza	Capitula, hydrorhiza	Capitula, hydrorhiza	Capitula, hydrorhiza, dactylozooid	Capitula, hydrorhiza, dactylozooid	Capitula, moniliform tentacles, body wall, hydrorhiza

Table 1. *Cont.*

Species	Zanclea divergens	Zanclea cf. protecta	Zanclea sp. 1	Zanclea sp. 2	Zanclea sango	Zanclea sp. (Clade I)	Asyncoryne ryniensis
Stenoteles size (µm)	11–15 × 9–12, 5–8 × 4–6	11–15 × 8–13, 5–8 × 4–7	14–17 × 12–13, 6–7 × 4–6	18–20 × 12–18, 15–16 × 12–15, 6–7 × 4–6	10–14 × 9–14, 6–8 × 5–6	10–14 × 8–13, 6–9 × 4–6	30–32 × 26–29, 10–11 × 7–8, 8–9 × 6–7
Euryteles	Macrobasic holotrichous, in hypostome, hydrorhiza	No	Macrobasic holotrichous, in hydrorhiza	Macrobasic holotrichous, in hydrorhiza	Macrobasic apotrichous, in hypostome, hydrorhiza, dactylozooid	Macrobasic apotrichous, in hypostome, hydrorhiza, dactylozooid	Macrobasic holotrichous, in body wall, hydrorhiza
Euryteles size (µm)	29–33 × 16–18	No	27–29 × 14–16	18–21 × 11–15	17–21 × 6–10	16–20 × 7–9	20–21 × 15–16

Table 2. Morphology of the newly released medusae.

Species	Zanclea divergens	Zanclea cf. protecta	Zanclea sp. 1	Zanclea sp. 2	Zanclea sango	Zanclea sp. (Clade I)	Asyncoryne ryniensis
Umbrella	Bell-shaped, diameter: 0.5 mm	Globular to bell-shaped, diameter: 0.5 mm	Globular, diameter: 0.15 mm	Globular, diameter: 0.2 mm	Bell-shaped, diameter: 0.7 mm	Bell-shaped, diameter: 0.5 mm	Bell-shaped, diameter: 0.5 mm
Canals	4 radials, one circular	4 radials, one circular	No	No	4 radials, one circular	4 radials, one circular	4 radials, one circular
Bulbs	4, 2 tentacular larger	4, 2 tentacular larger	2	2	4, 2 tentacular larger	4, 2 tentacular larger	4, 2 tentacular larger
Nematocyst pouches	4, 2 above tentacular bulbs larger	4, same size	No	No	4, same size	4, same size	4, same size

Table 2. *Cont.*

Species	*Zanclea divergens*	*Zanclea* cf. *protecta*	*Zanclea* sp. 1	*Zanclea* sp. 2	*Zanclea sango*	*Zanclea* sp. (Clade I)	*Asyncoryne ryniensis*
Manubrium	Cylindrical, 1/3 of the subumbrellar cavity	Cylindrical, 1/3 of the subumbrellar cavity	Reaching the velar opening, with 4 arms	Protruding from the bell cavity, with 4 arms	Cylindrical, 1/3 of the subumbrellar cavity	Cylindrical, 1/3 of the subumbrellar cavity	Cylindrical, 1/3 of the subumbrellar cavity
Tentacles	2	2	2	2	2	2	2
Cnidophores	21–31, slightly elongated	25–37, elongated	10–15, rounded to elongated	15–17, rounded	35–40, rounded	32–43, rounded	17–24
Color	Transparent, transparent to white manubrium	Transparent, whitish manubrium	Transparent, orange to white manubrium	Transparent, orange to whitish manubrium	Transparent, whitish manubrium	Transparent, whitish manubrium	Transparent, whitish to orange manubrium
Exumbrellar nematocysts (size in μm)	Isorhizae: 5–7 × 5–6	Basitrichous isorhizae: 5–6 × 4–5	Macrobasic holotrichous mastigophores: 7–8 × 6–7	Macrobasic holotrichous mastigophores: 7–9 × 6–8	No	No	Macrobasic holotrichous euryteles: 6–8 × 6–7
Pouches nematocysts (size in μm)	Stenoteles: 12–16 × 11–13	Stenoteles: 11–13 × 9–10	No	No	Macrobasic apotrichous euryteles: 18–19 × 7–9; Stenoteles: 11–12 × 10–11	Macrobasic apotrichous euryteles: 16–20 × 9–11; Stenoteles: 8–10 × 8–9	Stenoteles: 28–29 × 24–26
Manubrium nematocysts (size in μm)	Stenoteles: 5–8 × 4–6	Stenoteles: 6–7 × 5–6	No	No	Stenoteles: 8–10 × 7–8	Stenoteles: 7–8 × 5–6	No
Cnidophores nematocysts (size in μm)	Bean-shaped macrobasic holotrichous euryteles: 7–8 × 5–6	Bean-shaped macrobasic holotrichous euryteles: 7–8 × 5–6	Bean-shaped macrobasic apotrichous euryteles: 5–6 × 4–6	Bean-shaped macrobasic apotrichous euryteles: 6–8 × 4–5	Bean-shaped macrobasic apotrichous euryteles: 7–8 × 4–5	Bean-shaped macrobasic apotrichous euryteles: 7–8 × 4–5	Bean-shaped macrobasic euryteles: 7–8 × 6–7

3.2. Modifications of the Hosts

In all *Zanclea* samples, modification of the skeletons of the hosts were observed. *Zanclea divergens* polyps 'pierced' the skeleton of *Celleporaria vermiformis* along the border between zooids, and in some cases the bryozoan skeleton overgrew the base of polyps as a tube (Figure 1e). The hydrorhiza of *Zanclea* cf. *protecta* growing over the colony of bryozoan host *Parasmittina* cf. *spondylicola* was surrounded by a thin skeletal lamina produced exactly along the border between zooids (Figure 2d). Polyps of *Zanclea* sp. 1 and *Zanclea* sp. 2, associated with *Celleporaria pigmentaria* and *Celleporaria* sp. respectively, were observed coming out from the colony of the hosts at the borders between zooids, being partially overgrown at their base by the skeleton (Figure 3e, Figure 4e). Scleractinian-associated *Zanclea sango* and *Zanclea* sp. caused micro-alterations in the skeleton of the host corals, due to the skeletal overgrowth of the base of polyps and portions of the hydrorhiza (Figures 5f and 6f, respectively).

3.3. Green Fluorescence Essay

The six *Zanclea* and the *Asyncoryne* species showed different patterns of green fluorescence in both the polyp and medusa stages (Figure 8). Specifically, three *Zanclea* species (*Zanclea divergens*, *Zanclea* sp. 1, *Zanclea* sp. 2) and *Asyncoryne ryniensis* did not show fluorescence in the medusa stage. By contrast, the other three *Zanclea* species showed a marked green fluorescence in different structures. *Zanclea* cf. *protecta* showed a fluorescence at the level of the subumbrella, manubrium, and bulbs (Figure 8e,f). *Zanclea* sp. (Clade I) medusae released from colonies associated with *Goniastrea* sp. were characterized by a fluorescence of the radial and circular canals, bulbs, and whole manubrium (Figure 8a,b). Finally, *Zanclea sango* medusae displayed a pattern similar to that of *Zanclea* sp. (Clade I), with the exception of the central portion of the manubrium that did not show any fluorescence (Figure 8c,d). Fluorescence in these medusae was also present when still attached to the parental colony, and showed the same patterns displayed by newly released medusae (Figure 8g,h).

Regarding the polyp stages, *Zanclea* species did not show any fluorescence. Contrarily, *Asyncoryne ryniensis* polyps were characterized by a marked fluorescence at the base of moniliform tentacles (Figure 8i,j). In one polyp, green fluorescence was easily detected without excitation with blue light (Figure 7c).

Fluorescence patterns were identical for all medusae belonging to the same species, and no differences were detected between observations carried out at day one and five after release.

Fluorescence patterns of polyps and medusae for each species are summarized in Table 3.

Table 3. Summary of green fluorescence (GF) patterns in polyps and medusae of *Zanclea* and *Asyncoryne* species.

Species	Host/Substrate	Polyp GF	Medusa GF
Zanclea divergens	*Celleporaria vermiformis*	none	none
Zanclea cf. *protecta*	*Parasmittina* cf. *spondylicola; Schizoporella* sp.	none	Subumbrella, manubrium, bulbs
Zanclea sp. 1	*Celleporaria pigmentaria*	none	none
Zanclea sp. 2	*Celleporaria* sp.	none	none
Zanclea sango	*Pavona varians*	none	Manubrium (not in the middle), canals, bulbs
Zanclea sp. (Clade I)	*Goniastrea* sp.	none	Manubrium (whole), canals, bulbs
Asyncoryne ryniensis	Rock, sponge	base of tentacles	none

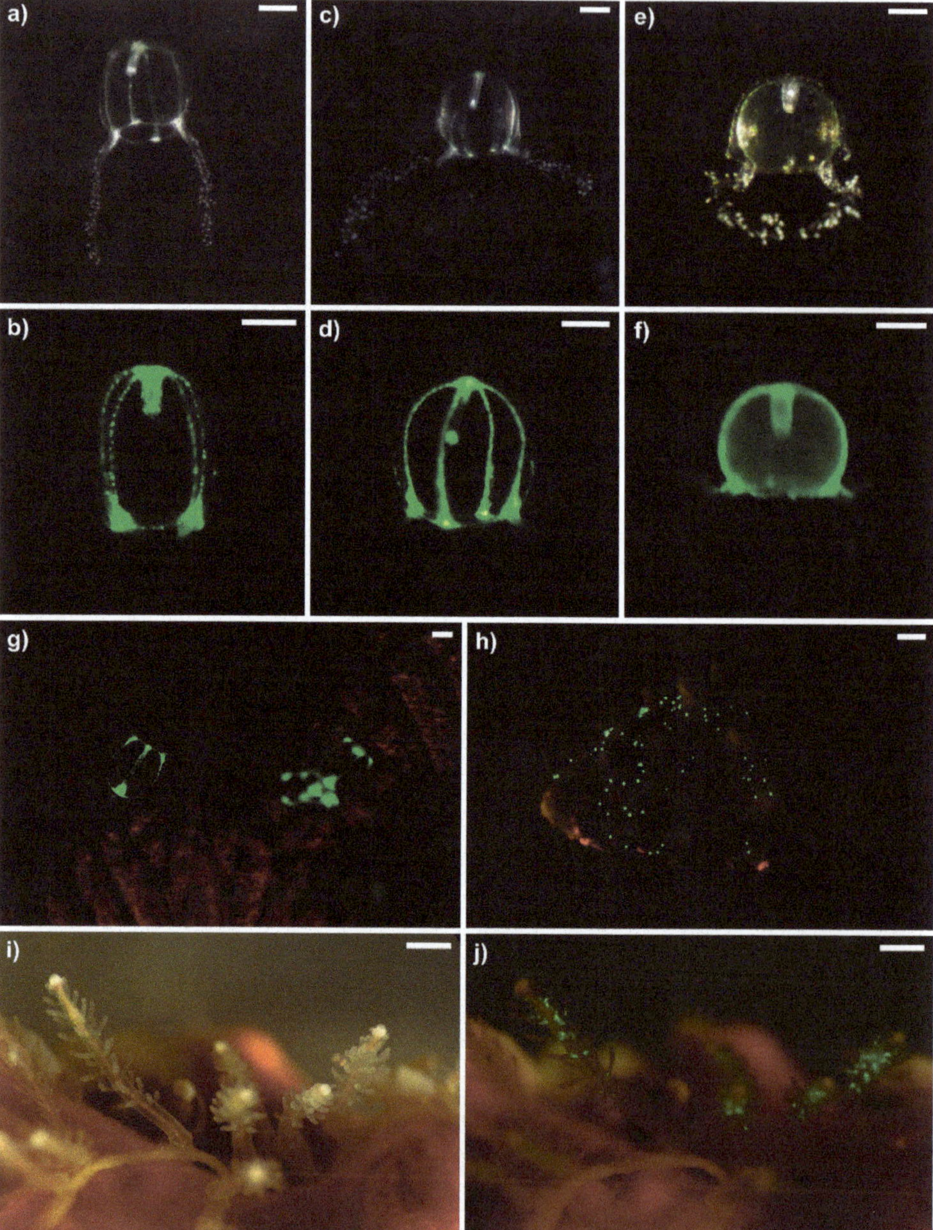

Figure 8. Green fluorescence in *Zanclea* and *Asyncoryne* species. (**a,b**) Medusa of *Zanclea* sp. (Clade I) released from a colony associated with *Goniastrea* sp.; (**c,d**) medusa of *Zanclea sango*; (**e,f**) medusa of *Zanclea protecta*; (**g**) medusa of *Zanclea* sp. before release, associated with *Goniastrea* sp.; (**h**) *Zanclea* cf. *protecta* medusa buds in the colony associated with *Schizoporella* sp. overgrowing the gastropod *Drupella* sp.; (**i,j**) *Asyncoryne ryniensis* polyps. Scale bars: (**a–g**) 0.2 mm; (**h**) 5 mm; (**i,j**) 1 mm.

3.4. 16S rRNA Phylogeny

DNA was extracted successfully, and *16S rRNA* sequences were generated for each analyzed sample. BLASTn searches resulted in a 100% match with previously deposited sequences obtained from Maldivian samples for *Zanclea* sp. 1, *Zanclea* sp. 2, *Zanclea* sp. (Clade I), and *Zanclea sango*. *Zanclea divergens* resulted in a match of 90.7% with an Indonesian sequence of the same species (MF000525), and this low value is explained by the fact that *Z. divergens* is a complex of cryptic species [31]. No *Zanclea protecta* sequences have been deposited so far, and the search for this species resulted in a match of 91.3% with *Zanclea costata* from the Mediterranean Sea (FN687559). Sequences of Maldivian *Asyncoryne ryniensis* resulted in a match of 98.4% with a Japanese specimen (EU876552).

The phylogenetic tree was rooted using *Pennaria disticha* [22,32] and, despite the overall poorly supported relationships (Figure 9), it agrees with previous reconstructions of *Zanclea* phylogeny [22]. Specifically, coral-associated *Zanclea* resulted in a fully supported clade, similarly to the clade composed of *Zanclea* sp. 1 and sp. 2 associated with bryozoans. Moreover, *Z. divergens* was well supported as the sister species of the latter clade, and all three species were associated with *Celleporaria* spp. Finally, the family Zancleidae was confirmed to be polyphyletic, due to the position of *Asyncoryne ryniensis*, which divides the family in two main clades, one associated with corals, and the other with bryozoans.

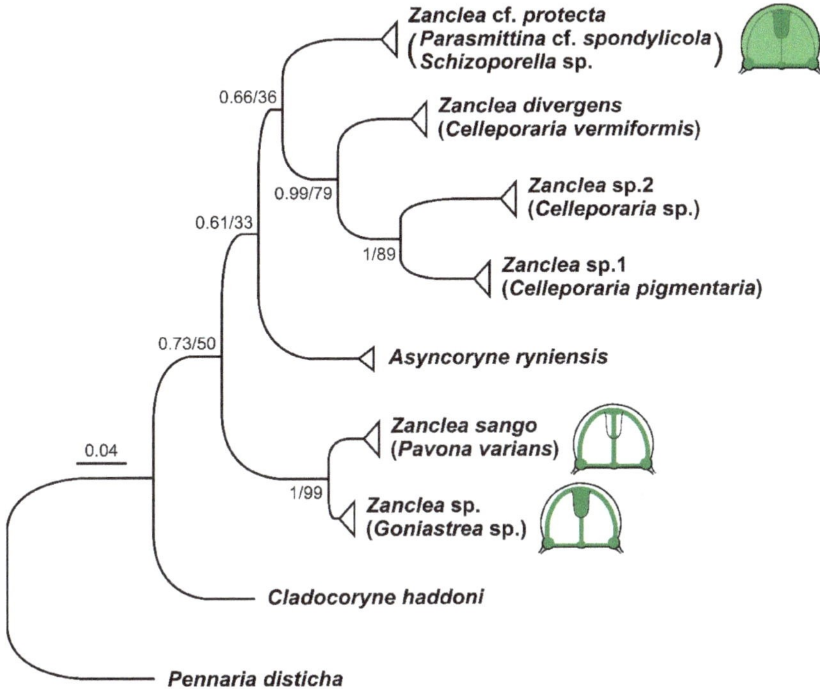

Figure 9. *16S rRNA* phylogeny of the species included in the analyses. Numbers at nodes represent Bayesian posterior probabilities and maximum likelihood bootstrap values, respectively. Hosts for each species are in brackets. Schematic drawings of fluorescence patterns in *Zanclea* medusae are also represented.

Inter-specific genetic distances were high in all comparisons, with the lowest level between the two coral-associated species *Z. sango* and *Zanclea* sp. (Clade I) (4%). All other species showed values higher than 10%. Intra-specific distances were equal to 0% in all cases (Table 4).

Table 4. Pairwise % uncorrected *p*-distances (*16S rRNA*) between all species analyzed.

	(1)	(2)	(3)	(4)	(5)	(6)	(7)
(1) *Z. divergens*	0						
(2) *Z. protecta*	11.7 (1.2)	0					
(3) *Zanclea* sp. 1	10.7 (1.3)	13.2 (1.4)	0				
(4) *Zanclea* sp. 2	12.5 (1.3)	13.7 (1.3)	10.7 (1.2)	0			
(5) *Z. sango*	12.9 (1.3)	13.0 (1.3)	14.9 (1.4)	14.9 (1.4)	0		
(6) *Zanclea* sp. (Clade I)	12.0 (1.3)	12.0 (1.3)	14.0 (1.4)	14.0 (1.3)	4.0 (0.8)	0	
(7) *A. ryniensis*	12.7 (1.3)	11.5 (1.3)	13.2 (1.4)	12.9 (1.3)	13.7 (1.4)	12.3 (1.3)	0

4. Discussion

The genus *Zanclea* and family Zancleidae are challenging taxa both from an evolutionary point of view and for species identification or description [19,22]. Indeed, the genus and family are polyphyletic [22,33], and further analyses are needed to establish new genera or even families. Their taxonomy is complicated by the fact that polyps often have intergrading morphologies, and the adult medusa must be observed and characterized for correct species identification and description [20,22]. Indeed, cryptic or unidentifiable species are common in the *Zanclea* genus [18,19,22].

In this work we analyzed the morphology of six *Zanclea* species, considering the general features of polyps and medusae, the cnidome of both life stages, the alteration of the host skeletal structures, and the green fluorescence patterns. Additionally, we analyzed the molecular identity, phylogenetic relationships and genetic diversity of the species, confirming their possible belonging to six well separated *Zanclea* lineages. Our results show that the characterization of general morphology, and cnidome is in some cases enough to distinguish between *Zanclea* species. For instance, by combining observations on the presence, localization and size of euryteles, and the general appearance of polyps and medusae, it is possible to distinguish the analyzed bryozoan-associated species. By contrast, scleractinian-associated species showed a very similar morphology, as already documented in previous studies [18,21,34].

In all *Zanclea* species here analyzed, alterations of the host skeleton were observed. The bryozoan *Parasimittina* cf. *spondylicola* showed the most evident modification, with the skeletal lamina overgrowing the hydrorhiza of *Zanclea* cf. *protecta*, as already noted by Hasting [35] and Boero et al. [20] for *Zanclea protecta* associated with *Parasmittina crosslandi* (Hastings, 1930) and other unidentified bryozoans. A similar situation was observed for *Celleporaria–Zanclea* associations, where the base of polyps was occasionally surrounded by bryozoan skeletal structures. Additionally, scleractinians hosting *Zanclea* showed micro-alterations related to the presence of symbionts, as already observed in *Goniastrea*, *Pavona*, and *Porites* corals [21]. The presence of these modifications may support the hypothesis that at least some *Zanclea* species are mutualistically associated with their hosts, since they may provide additional protection and competitive advantages to their hosts and in turn benefit from being partially enclosed in hard carbonatic structures [36,37].

Differences were found in the green fluorescence patterns of *Zanclea* and *Asyncoryne* species. *Zanclea divergens*, *Zanclea* sp. 1 and *Zanclea* sp. 2 did not show any fluorescence neither in the polyps nor in the medusae. *Zanclea* cf. *protecta*, *Zanclea sango*, and *Zanclea* sp. (Clade I) did not show any fluorescence in the polyps but medusae were characterized by different green fluorescence patterns. Finally, *Asyncoryne ryniensis*, which has different polyps but medusae very similar to those of *Zanclea*, showed fluorescence in polyps but not in medusae. *Zanclea* cf. *protecta* is characterized by a diffuse fluorescence in bulbs, manubrium, and subumbrella, whereas *Z. sango* and *Zanclea* sp. (Clade I) are fluorescent in bulbs, manubrium, and canals. Despite the two latter coral-associated species have overlapping morphologies in both polyp and medusa stages, medusae showed differences in the distribution of green fluorescent proteins at the level of manubrium. Specifically, in *Zanclea* sp. (Clade I) the entire manubrium is fluorescent, and this pattern is visible even in the medusa buds still attached to the parental colony, whereas in *Zanclea sango* fluorescence is concentrated at the extremes of manubrium (mouth and close to the umbrellar margin), being absent in the middle portion. These conditions were observed in all analyzed medusae and therefore may be taxonomically informative, even if further analyses are needed to confirm this hypothesis. Fontana et al. [38] also found green fluorescence in medusa buds of *Acropora*-associated *Zanclea* species, but the localization was not reported. However, this suggests that, potentially, the medusae of other coral-associated *Zanclea* species may be fluorescent. If this is true, the investigation of fluorescence patterns in all *Zanclea* species associated with scleractinians may help disentangling the cryptic diversity that characterize this group.

The function, if any, of green fluorescent proteins in the analyzed species is still not clear. One of the possible explanations is attraction of prey [8]. The polyps of the six *Zanclea* species observed all live in symbiosis with other organisms, and the lack of fluorescence in this stage may be related to specific feeding interactions with the hosts, as described for *Zanclea divergens*, which seems to feed on mucous aggregates of particles egested by the bryozoan [39]. Moreover, *Asyncoryne ryniensis* is not symbiotic, and fluorescence is found at the base of polyp tentacles. This explanation complicates with the medusa stages, since species with potentially similar feeding behaviors show contrasting fluorescence patterns.

Overall, the results obtained in this work show that the combination of multiple approaches allows one to discriminate closely related *Zanclea* species and provide information on the relationships between these hydrozoans and their hosts. Additionally, the analysis of green fluorescence patterns seems to be a promising tool for hydrozoan taxonomy and should be performed at a large scale to assess its adequacy in exploring and distinguishing the diversity of enigmatic hydrozoan taxa, such as zancleids.

Author Contributions: Conceptualization, D.M. and S.M.; investigation, D.M., L.S., A.S., and A.N.O.; writing—original draft preparation, all authors.; writing—review and editing, all authors. All authors have read and agreed to the published version of the manuscript.

Funding: This research received no external funding.

Acknowledgments: A. Ostrovsky thanks the Austrian Science Fund (project P19337-B17) for supporting the taxonomical research on the tropical Bryozoa. The authors are grateful to two anonymous reviewers, whose comments greatly improved the manuscript.

Conflicts of Interest: The authors declare no conflict of interest.

References

1. Chudakov, D.M.; Matz, M.V.; Lukyanov, S.; Lukyanov, K.A. Fluorescent proteins and their applications in imaging living cells and tissues. *Physiol. Rev.* **2010**, *90*, 1103–1163. [CrossRef]
2. Shimomura, O.; Johnson, F.H.; Saiga, Y. Extraction, purification and properties of aequorin, a bioluminescent protein from the luminous hydromedusan, *Aequorea. J. Cell. Comp. Physiol.* **1962**, *59*, 223–239. [CrossRef]
3. Miyawaki, A. Green fluorescent protein-like proteins in reef Anthozoa animals. *Cell Struct. Funct.* **2002**, *27*, 343–347. [CrossRef]
4. Shagin, D.A.; Barsova, E.V.; Yanushevich, Y.G.; Fradkov, A.F.; Lukyanov, K.A.; Labas, Y.A.; Semenova, T.N.; Ugalde, J.A.; Meyers, A.; Nunez, J.M.; et al. GFP-like proteins as ubiquitous metazoan superfamily: evolution of functional features and structural complexity. *Mol. Biol. Evol.* **2004**, *21*, 841–850. [CrossRef]
5. Salih, A.; Larkum, A.; Cox, G.; Kühl, M.; Hoegh-Guldberg, O. Fluorescent pigments in corals are photoprotective. *Nature* **2000**, *408*, 850. [CrossRef]
6. Morin, J.G.; Hastings, J.W. Energy transfer in a bioluminescent system. *J. Cell. Physiol.* **1971**, *77*, 313–318. [CrossRef]
7. Sparks, J.S.; Schelly, R.C.; Smith, W.L.; Davis, M.P.; Tchernov, D.; Pieribone, V.A.; Gruber, D.F. The covert world of fish biofluorescence: A phylogenetically widespread and phenotypically variable phenomenon. *PLoS ONE* **2014**, *9*, e83259. [CrossRef]
8. Haddock, S.H.; Dunn, C.W. Fluorescent proteins function as a prey attractant: experimental evidence from the hydromedusa *Olindias formosus* and other marine organisms. *Biol. Open* **2015**, *4*, 1094–1104. [CrossRef] [PubMed]
9. De Brauwer, M.; Hobbs, J.P.A. Stars and stripes: biofluorescent lures in the striated frogfish indicate role in aggressive mimicry. *Coral Reefs* **2016**, *35*, 1171. [CrossRef]

10. Prudkovsky, A.A.; Ivanenko, V.N.; Nikitin, M.A.; Lukyanov, K.A.; Belousova, A.; Reimer, J.D.; Berumen, M.L. Green fluorescence of *Cytaeis* hydroids living in association with *Nassarius* gastropods in the Red Sea. *PLoS ONE* **2016**, *11*, e0146861. [CrossRef] [PubMed]

11. Kubota, S.; Pagliara, P.; Gravili, C. Fluorescence distribution pattern allows to distinguish two species of *Eugymnanthea* (Leptomedusae: Eirenidae). *J. Mar. Biol. Assoc. U. K.* **2008**, *88*, 1743–1746. [CrossRef]

12. Kubota, S. Various distribution patterns of green fluorescence in small hydromedusae. *Kuroshio Biosphere* **2010**, *6*, 11–14.

13. Kubota, S.; Gravili, C. Rare distribution of green fluorescent protein (GFP) in hydroids from Porto Cesareo, Lecce, Italy, with reference to biological meaning of this rarity. *Biogeogr.* **2011**, *13*, 9–11.

14. Ryusaku, D.; Mayu, O.; Hiroshi, N. Green fluorescent protein (GFP)-like substance in the hydrozoan jellyfish *Cytaeis uchidae*: examination of timing and localization of its expression and utilization for biological education. *Bull. Miyagi Univ. Educ.* **2012**, *47*, 95–100.

15. Maggioni, D.; Montano, S.; Seveso, D.; Galli, P. Molecular evidence for cryptic species in *Pteroclava krempfi* (Hydrozoa, Cladocorynidae) living in association with alcyonaceans. *Syst. Biodivers.* **2016**, *14*, 484–493. [CrossRef]

16. Montano, S.; Maggioni, D.; Galli, P.; Hoeksema, B.W. A cryptic species in the *Pteroclava krempfi* species complex (Hydrozoa, Cladocorynidae) revealed in the Caribbean. *Mar. Biodivers.* **2017**, *47*, 83–89. [CrossRef]

17. Miglietta, M.P.; Maggioni, D.; Matsumoto, Y. Phylogenetics and species delimitation of two hydrozoa (phylum Cnidaria): *Turritopsis* (McCrady, 1857) and *Pennaria* (Goldfuss, 1820). *Mar. Biodivers.* **2019**, *49*, 1085–1100. [CrossRef]

18. Montano, S.; Maggioni, D.; Arrigoni, R.; Seveso, D.; Puce, S.; Galli, P. The hidden diversity of *Zanclea* associated with scleractinians revealed by molecular data. *PLoS ONE* **2015**, *10*, e0133084. [CrossRef]

19. Maggioni, D.; Montano, S.; Arrigoni, R.; Galli, P.; Puce, S.; Pica, D.; Berumen, M.L. Genetic diversity of the *Acropora*-associated hydrozoans: new insight from the Red Sea. *Mar. Biodivers.* **2017**, *47*, 1045–1055. [CrossRef]

20. Boero, F.; Bouillon, J.; Gravili, C. A survey of *Zanclea*, *Halocoryne* and *Zanclella* (Cnidaria, Hydrozoa, Anthomedusae, Zancleidae) with description of new species. *Ital. J. Zool.* **2000**, *67*, 93–124. [CrossRef]

21. Manca, F.; Puce, S.; Caragnano, A.; Maggioni, D.; Pica, D.; Seveso, D.; Galli, P.; Montano, S. Symbiont footprints highlight the diversity of scleractinia - associated *Zanclea* hydrozoans (Cnidaria, Hydrozoa). *Zool. Scr.* **2019**, *48*, 399–410. [CrossRef]

22. Maggioni, D.; Arrigoni, R.; Galli, P.; Berumen, M.L.; Seveso, D.; Montano, S. Polyphyly of the genus *Zanclea* and family Zancleidae (Hydrozoa, Capitata) revealed by the integrative analysis of two bryozoan-associated species. *Contrib. Zool.* **2018**, *87*, 87–104. [CrossRef]

23. Zietara, M.S.; Arndt, A.; Geets, A.; Hellemans, B.; Volckaert, F.A. The nuclear rDNA region of *Gyrodactylus arcuatus* and *G. branchicus* (Monogenea: Gyrodactylidae). *J. Parasitol.* **2000**, *86*, 1368–1373. [CrossRef]

24. Maggioni, D.; Puce, S.; Galli, P.; Seveso, D.; Montano, S. Description of *Turritopsoides marhei* sp. nov. (Hydrozoa, Anthoathecata) from the Maldives and its phylogenetic position. *Mar. Biol. Res.* **2017**, *13*, 983–992. [CrossRef]

25. Cunningham, C.W.; Buss, L.W. Molecular evidence for multiple episodes of paedomorphosis in the family Hydractiniidae. *Biochem. Syst. Ecol.* **1993**, *21*, 57–69. [CrossRef]

26. Katoh, K.; Standley, D.M. MAFFT multiple sequence alignment software version 7: improvements in performance and usability. *Mol. Biol. Evol.* **2013**, *30*, 772–780. [CrossRef]

27. Darriba, D.; Taboada, G.L.; Doallo, R.; Posada, D. jModelTest 2: more models, new heuristics and parallel computing. *Nat. Methods* **2012**, *9*, 772. [CrossRef]

28. Ronquist, F.; Teslenko, M.; van der Mark, P.; Ayres, D.L.; Darling, A.; Höhna, S.; Larget, B.; Liu, L.; Suchard, M.A.; Huelsenbeck, J.P. MrBayes 3.2: efficient Bayesian phylogenetic inference and model choice across a large model space. *Syst. Biol.* **2012**, *61*, 539–542. [CrossRef] [PubMed]

29. Stamatakis, A. RAxML version 8: a tool for phylogenetic analysis and post-analysis of large phylogenies. *Bioinformatics* **2014**, *30*, 1312–1313. [CrossRef] [PubMed]

30. Kumar, S.; Stecher, G.; Li, M.; Knyaz, C.; Tamura, K. MEGA X: molecular evolutionary genetics analysis across computing platforms. *Mol. Biol. Evol.* **2018**, *35*, 1547–1549. [CrossRef] [PubMed]

31. Maggioni, D. (Universoty of Milano-Bicocca). Personal Observation. 2020.

32. Maggioni, D.; Galli, P.; Berumen, M.L.; Arrigoni, R.; Seveso, D.; Montano, S. *Astrocoryne cabela*, gen. nov. et sp. nov. (Hydrozoa: Sphaerocorynidae), a new sponge-associated hydrozoan. *Invertebr. Syst.* **2017**, *31*, 734–746. [CrossRef]

33. Nawrocki, A.M.; Schuchert, P.; Cartwright, P. Phylogenetics and evolution of Capitata (Cnidaria: Hydrozoa), and the systematics of Corynidae. *Zool. Scr.* **2010**, *39*, 290–304. [CrossRef]

34. Montano, S.; Arrigoni, R.; Pica, D.; Maggioni, D.; Puce, S. New insights into the symbiosis between *Zanclea* (Cnidaria, Hydrozoa) and scleractinians. *Zool. Scr.* **2015**, *44*, 92–105. [CrossRef]

35. Hastings, A.B. LIV. On the association of a gymnoblastic Hydroid (*Zanclea protecta*, sp. n.) with various cheilostomatous polyzoa from the Tropical E. Pacific. *Ann. Mag. Nat. Hist.* **1930**, *5*, 552–560. [CrossRef]

36. Montano, S.; Fattorini, S.; Parravicini, V.; Berumen, M.L.; Galli, P.; Maggioni, D.; Arrigoni, R.; Seveso, D.; Strona, G. Corals hosting symbiotic hydrozoans are less susceptible to predation and disease. *Proc. R. Soc. B* **2017**, *284*, 20172405. [CrossRef]

37. Osman, R.W.; Haugsness, J.A. Mutualism among sessile invertebrates: a mediator of competition and predation. *Science* **1981**, *211*, 846–848. [CrossRef]

38. Fontana, S.; Keshavmurthy, S.; Hsieh, H.J.; Denis, V.; Kuo, C.Y.; Hsu, C.M.; Leung, J.K.L.; Tsai, W.S.; Wallace, C.C.; Chen, C.A. Molecular evidence shows low species diversity of coral-associated hydroids in *Acropora* corals. *PLoS ONE* **2012**, *7*, e50130. [CrossRef]

39. Puce, S.; Cerrano, C.; Boyer, M.; Ferretti, C.; Bavestrello, G. *Zanclea* (Cnidaria: Hydrozoa) species from Bunaken Marine Park (Sulawesi Sea, Indonesia). *J. Mar. Biol. Assoc. U. K.* **2002**, *82*, 943–954. [CrossRef]

Interesting Images

Extension of the Recorded Host Range of Caribbean Christmas Tree Worms (*Spirobranchus* spp.) with Two Scleractinians, a Zoantharian, and an Ascidian

Bert W. Hoeksema [1,2,3,*], **Jaaziel E. García-Hernández** [4], **Godfried W.N.M. van Moorsel** [5,6], **Gabriël Olthof** [1,3] and **Harry A. ten Hove** [1]

[1] Taxonomy and Systematics Group, Naturalis Biodiversity Center, P.O. Box 9517, 2300 RA Leiden, The Netherlands; gabriel.olthof@gmail.com (G.O.); harry.tenhove@naturalis.nl (H.A.t.H.)
[2] Groningen Institute for Evolutionary Life Sciences, University of Groningen, P.O. Box 11103, 9700 CC Groningen, The Netherlands
[3] Institute of Biology Leiden, Leiden University, P.O. Box 9505, 2300 RA Leiden, The Netherlands
[4] Marine Genomic Biodiversity Laboratory, University of Puerto Rico - Mayagüez, La Parguera, Puerto Rico 00667, USA; jaaziel.garcia@upr.edu
[5] Ecosub, Berkenlaantje 2, 3956 DM Leersum, The Netherlands; vanmoorsel@ecosub.nl
[6] ANEMOON Foundation, P.O. Box 29, 2120 AA Bennekom, The Netherlands
[*] Correspondence: bert.hoeksema@naturalis.nl; Tel.: +31-71-7519-631

Received: 5 March 2020; Accepted: 20 March 2020; Published: 21 March 2020

Caribbean Christmas tree worms (Annelida: Polychaeta: Serpulidae: *Spirobranchus*) are considered host generalists in their associations with anthozoan (Scleractinia) and hydrozoan (*Millepora*) stony corals [1–4]. As planktonic larvae, they settle on coral surfaces and start secreting a calcareous tube to be used as a dwelling. This tube usually becomes overgrown by the host coral (except for its opening) and may get encapsulated deep inside the coral skeleton. In this manner, the well-protected worms grow and survive predation [5] and other hazards, allowing them to live for over four decades [6]. When the host corals are overgrown by other organisms, such as octocorals and sponges, these may act as secondary hosts [7,8].

The long lists of Caribbean host species suggest that the recorded number has reached a maximum [1–4]. However, recent surveys (2015–2019) in the southern and eastern Caribbean, as well as in the Greater Antilles, enabled us to establish new records of two primary hosts (scleractinians) and two secondary hosts (a zoantharian and an ascidian).

The coral–worm associations occurred in shallow subtidal water (<4 m depth), with *Pseudodiploria clivosa* (Ellis & Solander, 1786) hosting *Spirobranchus giganteus* (Pallas, 1766) at St. Eustatius (Figure 1) and *Favia fragum* (Esper, 1795) hosting both *S. giganteus* (Figure 2a–c) and *S. polycerus* (Schmarda, 1861) (Figure 2d) at Bonaire. The secondary host observations, both for *S. giganteus*, involved the zoantharian *Palythoa caribaeorum* (Duchassaing & Michelotti, 1860) at Puerto Rico (Figure 3) and the ascidian *Trididemnum solidum* (Van Name, 1902) at Bonaire and Curaçao (Figure 4). *Palythoa caribaeorum* represents a first record as a secondary host for a species of the order Zoantharia. Until now, the only other anthozoan secondary hosts were species in the order Alcyonacea (subclass Octorallia) [7], whereas *T. solidum* represents an entirely new host phylum, viz. Chordata. The only other non-anthozoan secondary hosts known to date are sponges (Porifera) [8].

The two new scleractinian hosts are both typical for shallow subtidal water near the shoreline (<4 m depth), where a lack of previous surveys may explain why they have not previously been reported. The new records of secondary hosts are remarkable because these encrusting animals are known to be aggressive in competition for space with scleractinians by allelopathy [9,10] and can be abundant on shallow reef flats and slopes, where they usually outcompete and kill scleractinian corals by overgrowing them [9,10]. In both cases, the Christmas tree worms survive by withstanding this overgrowth and maintain an open space near the tube opening (Figure 3d, Figure 4b,e).

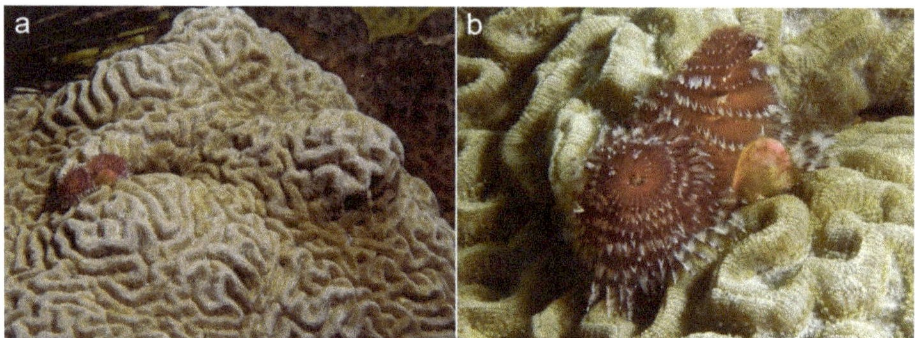

Figure 1. A coral of *Pseudodiploria clivosa* at 2 m depth, Scubaqua house reef (17°28′56″ N 62°59′20″ W), St. Eustatius, Eastern Caribbean (2015) hosting *Spirobranchus giganteus*: (**a**) overall view and (**b**) close-up.

Figure 2. *Favia fragum* hosting *Spirobranchus* spp. at 3–4 m depth, dive site "Front Porch" (12°09′ 54″ N 68°17′12″ W), Bonaire, Southern Caribbean (2019). (**a–c**) *Spirobranchus giganteus*: overall view (**a**), overgrown tube section indicated by red arrow (**b**); antler-shaped opercular spines showing dark pink coloration indicated by yellow arrow (**c**). (**d**) *Spirobranchus polycerus*: two individuals, one showing white spines on its operculum (blue arrow).

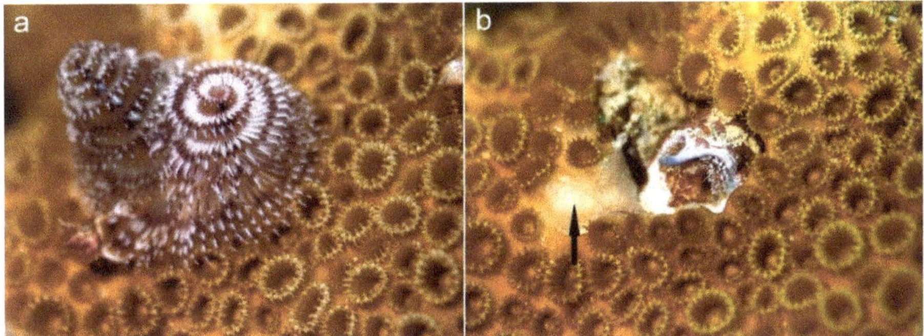

Figure 3. *Palythoa caribaeorum* acting as a secondary host for *Spirobranchus giganteus* at 5 m depth, Cayo Media Luna (La Parguera Natural Reserve), Puerto Rico, Greater Antilles (2017): (**a**) worm extended and (**b**) retracted, showing the tube opening surrounded by dead coral; damage to the zoantharian host caused by the operculum of the extended worm indicated by a black arrow.

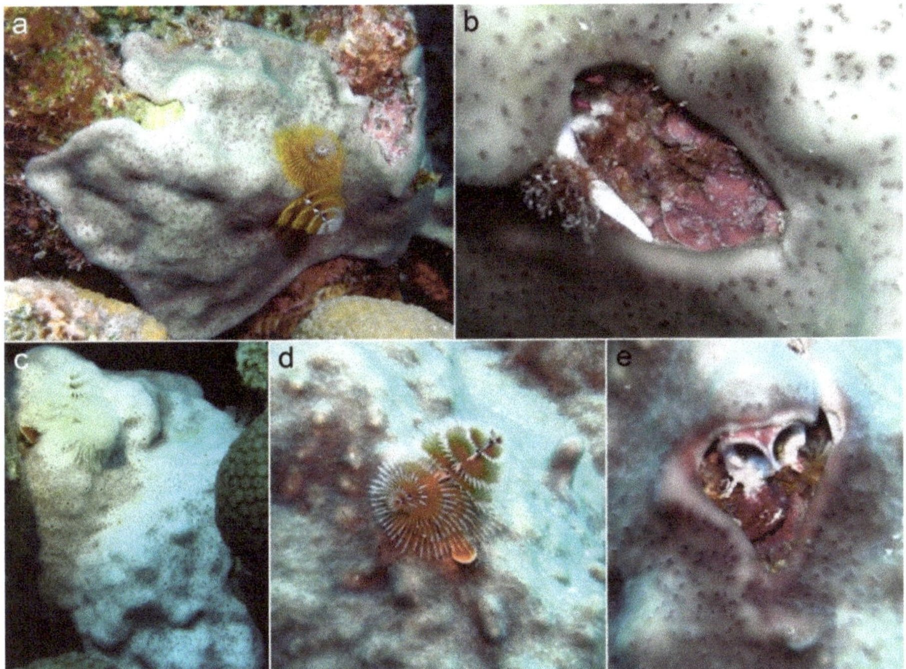

Figure 4. *Trididemnum solidum* acting as a secondary host for *Spirobranchus giganteus* in the Southern Caribbean: (**a,b**) dive site "Thousand Steps" (12°12′39″ N 68°19′17″ W), Bonaire (2019); (**c**) Marie Pampoen (12°05′31″ N 68°54′27″ W), Curaçao, 12 m depth (2017); (**d,e**) Daaibooi Bay (12°12′41″ N 69°05′13″ W), Curaçao (2017). Extended worms (**a,d**) and the same individuals retracted, showing an open space in front of the worm tube mouth (**b,e**).

Our new host records confirm two Caribbean Christmas tree worms as generalist symbionts capable of infesting a large spectrum of host corals. They are also strong survivors when their primary hosts become overgrown by more aggressive competitors for space. Previous host records mostly concern *S. giganteus* [1–4], but here we also report a new host coral for *S. polycerus*. This worm species occurs in shallow water (<4 m depth) [5], whereas *S. giganteus* is commonly found down to

40 m depth [1]. Both *Spirobranchus* species can easily be distinguished [5], as *S. giganteus* shows long dark pink opercular spines (Figure 2c), whereas those of *S. polycerus* are short and white (Figure 2d). Furthermore, *S. giganteus* may be larger than *S. polycerus* [5] and usually shows six to seven (maximum eight) whorls in its branchial spires, whereas *S. polycerus* has two to three (maximum five) [11]. Our observations suggest that future surveys may discover other hosts for both *Spirobranchus* species with the possibility of more host overlap. Whether such host sharing is related to their phylogenetic affinities or to ecological similarities (e.g., overlapping bathymetric distributions) is an open question that merits assessment.

Author Contributions: Conceptualization and supervision, B.W.H.; methodology, illustrations and funding acquisition, B.W.H., J.E.G.-H., G.W.N.M.v.M., and G.O., investigation, B.W.H., J.E.G.-H., G.W.N.M.v.M., G.O. and H.A.t.H.; writing—original draft preparation, B.W.H.; writing—review and editing, B.W.H., J.E.G.-H., G.W.N.M.v.M., G.O. and H.A.t.H. All authors have read and agreed to the published version of the manuscript.

Funding: Fieldwork at Bonaire was supported by the WWF Netherlands Biodiversity Fund, the Treub Maatschappij - Society for the Advancement of Research in the Tropics, and by the Nature of the Netherlands program of Naturalis Biodiversity Center.

Acknowledgments: We thank the staff of CARMABI (Curaçao) and CNSI (St. Eustatius) for their hospitality and assistance during the fieldwork. We thank STENAPA and Scubaqua Dive Center (Sint Eustatius), the Dive Shop (Curaçao), STINAPA, DCNA and Dive Friends (Bonaire) for logistic support. We are grateful to two anonymous reviewers for their constructive remarks on the manuscript.

Conflicts of Interest: The authors declare no conflict of interest. The funders had no role in the design of the study; in the collection, analyses, or interpretation of data; in the writing of the manuscript, or in the decision to publish the results.

References

1. Hoeksema, B.W.; ten Hove, H.A. The invasive sun coral *Tubastraea coccinea* hosting a native Christmas tree worm at Curaçao, Dutch Caribbean. *Mar. Biodivers.* **2017**, *47*, 59–65. [CrossRef]

2. Hoeksema, B.W.; van Beusekom, M.; ten Hove, H.A.; Ivanenko, V.N.; van der Meij, S.E.T.; van Moorsel, G.W.N.M. *Helioseris cucullata* as a host coral at St. Eustatius, Dutch Caribbean. *Mar. Biodivers.* **2017**, *47*, 71–78. [CrossRef]

3. Martin, D.; Britayev, T.A. Symbiotic polychaetes revisited: An update of the known species and relationships (1998–2017). *Oceanogr. Mar. Biol. Ann. Rev.* **2018**, *56*, 371–448.

4. Hoeksema, B.W.; Wels, D.; van der Schoot, R.J.; ten Hove, H.A. Coral injuries caused by *Spirobranchus* opercula with and without epibiotic turf algae at Curaçao. *Mar. Biol.* **2019**, *166*, 60. [CrossRef]

5. Hoeksema, B.W.; ten Hove, H.A. Attack on a Christmas tree worm by a Caribbean sharpnose pufferfish at St. Eustatius, Dutch Caribbean. *Bull. Mar. Sci.* **2017**, *93*, 1023–1024. [CrossRef]

6. Nishi, E.; Nishihira, M. Age-estimation of the Christmas tree worm *Spirobranchus giganteus* (Polychaeta, Serpulidae) living buried in the coral skeleton from the coral-growth band of the host coral. *Fish. Sci.* **1996**, *62*, 400–403. [CrossRef]

7. Hoeksema, B.W.; Lau, Y.W.; ten Hove, H.A. Octocorals as secondary hosts for Christmas tree worms off Curaçao. *Bull. Mar. Sci.* **2015**, *91*, 489–490. [CrossRef]

8. García-Hernández, J.E.; Hoeksema, B.W. Sponges as secondary hosts for Christmas tree worms at Curaçao. *Coral Reefs* **2017**, *36*, 1243. [CrossRef]

9. Suchanek, T.H.; Green, D.J. Interspecific competition between *Palythoa caribaeorum* and other sessile invertebrates on St. Croix reefs, US Virgin Islands. *Proc. 4th Int. Coral Reef Symp.* **1981**, *2*, 679–684.

10. Bak, R.P.M.; Sybesma, J.; van Duyl, F.C. The ecology of the tropical compound ascidian *Trididemnum solidum*. II. Abundance, growth and survival. *Mar. Ecol. Prog. Ser.* **1981**, *6*, 43–52. [CrossRef]

11. ten Hove, H.A. Serpulinae (Polychaeta) from the Caribbean: I–The genus *Spirobranchus*. *Stud. Fauna Curaçao Caribb. Is.* **1970**, *32*, 1–57.

 *diversity*

Interesting Images

Host-related Morphological Variation of Dwellings Inhabited by the Crab *Domecia acanthophora* in the Corals *Acropora palmata* and *Millepora complanata* (Southern Caribbean)

Bert W. Hoeksema [1,2,3,*] **and Jaaziel E. García-Hernández** [4]

1   Taxonomy and Systematics Group, Naturalis Biodiversity Center, 2300 RA Leiden, The Netherlands
2   Groningen Institute for Evolutionary Life Sciences, University of Groningen,
    9700 CC Groningen, The Netherlands
3   Institute of Biology Leiden, Leiden University, 2300 RA Leiden, The Netherlands
4   Marine Genomic Biodiversity Laboratory, University of Puerto Rico-Mayagüez, La Parguera, PR 00667, USA;
    jaaziel.garcia@upr.edu
*   Correspondence: bert.hoeksema@naturalis.nl; Tel.: +31-717-519-6311

Received: 21 March 2020; Accepted: 3 April 2020; Published: 5 April 2020

Brachyuran crabs of various families are known as obligate associates of stony corals, with many of these species living as endosymbionts inside the skeleton of their hosts [1]. In particular, coral gall crabs (Cryptochiridae) have been well studied in tropical coral reefs around the world. These crabs can be recognized by the shape of their dwellings (or pits), which may be crescent-shaped or resemble a slit, a canopy, a basket, or a gall, depending on the identity and morphology of their host, and on the position inside the host's skeleton [2–7]. Cryptochirids are each known to be associated with a few scleractinian host species (Anthozoa: Scleractinia) or only one [2–7]. Crabs of the species *Latopilumnus tubicolus* Türkay and Schuhmacher, 1985 (Pilumnidae), have so far only been reported as endosymbionts of the Indo-Pacific scleractinian *Tubastraea micranthus* (Ehrenberg, 1834) [8]. Their dwellings are unique because they start in one of the coral's calyces from where they penetrate deep inside the coral branches [8], becoming long and tubular, whereas the pits (or cysts) of cryptochirids remain relatively shallow [2–7].

Other records of crabs living inside stony corals, concern unidentified species of the genera *Tetralia* Dana, 1851 (Tetraliidae) and *Cymo* De Haan, 1833 (Xanthidae) that live in association with Indo-Pacific *Acropora* spp. [9]. The pits constructed by *Cymo* sp. show an oval margin that becomes increasingly thick, resembling a collar on top of flattened, fused *Acropora* branches, whereas those of *Tetralia* sp. resemble a slit in between bifurcating, round *Acropora* branches, which eventually develop into galls [9]. Dwellings made by Atlantic *Platypodiella* spp. (Xanthidae) are usually oval but can start as shallow depressions inside hosts of the genus *Palythoa* (Anthozoa: Zoantharia) [10,11], which are encrusting, colonial sea anemones related to scleractinians; they are leathery in appearance with sand particles inside their tissue instead of a calcareous skeleton.

The pits made by *Cymo* sp. resemble most those of the Atlantic crab *Domecia acanthophora* (Desbonne and Schramm, 1867) (Domeciidae) in flat branches of the scleractinian *Acropora palmata* (Lamarck, 1816) observed at Puerto Rico (Greater Antilles) and Venezuela (Southern Caribbean) [12,13]. Here, the same crab was also found in *Acropora cervicornis* (Lamarck, 1816) and *Acropora prolifera* (Lamarck, 1816), while it was also observed wandering on corals of other species [12,13]. *Domecia acanthophora* has also been reported as associated fauna of fire corals, like *Millepora alcicornis* Linneaus, 1758 at Brazil [14] and *Millepora* spp, at Yucatán, México [15]. However, no information is available on the morphology of *Domecia* pits in *Millepora* and its possible difference with *Acropora*. Therefore, we present information on *Domecia* dwellings found in six corals of *A. palmata* at Curaçao (<2 m depth;

2015) and two corals of *Millepora complanata* Lamarck, 1816 at Klein Bonaire (2–3 m depth; 2019). Curaçao and Bonaire (next to Klein Bonaire) are major Dutch Caribbean islands, situated 50 km apart and 70–80 km off the coast of Venezuela. The observations were made during biodiversity surveys down to 30 m depth using the roving diving technique.

The *Domecia* pits in *A. palmata* were up to 2 cm long (Figure 1). Those at the margins of coral branches had outlines that were not fully closed (Figure 1d,e), whereas pits away from the margins were enclosed and showed a thickened, smooth periphery, like a collar around an open callus (Figure 1h–i). An intermediate form consisted of a fully enclosed pit without a thickened margin (Figure 1f). The crabs could still be observed when they were inside the dwelling (Figure 1d–i) and also could easily be collected from an incomplete pit at a branch margin (Figure 1b).

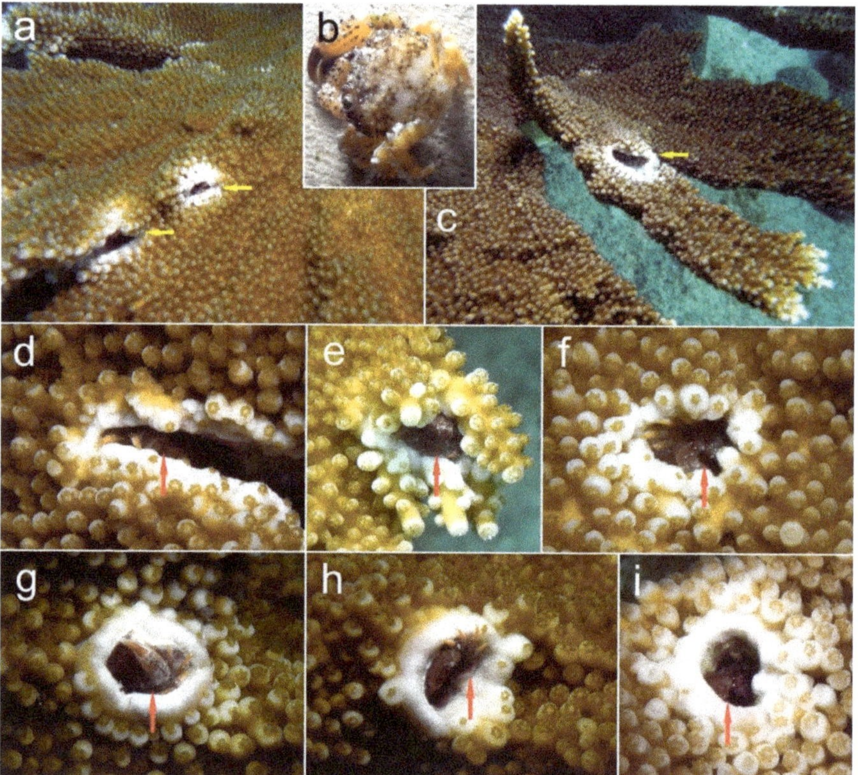

Figure 1. *Acropora palmata* at <2 m depth, Blue Wall dive site (2015), Curaçao (12°08′06″ N, 68°59′16″ W), hosting *Domecia acanthophora*: (**a,c**) branches with crab pits; (**b**) a crab taken from a pit; (**d–e**) crab dwellings at the margin of a coral branch; and (**f–i**) crab pits away from the branch margin, showing a white collar. Yellow arrows: crab dwellings; red arrows: Crabs.

The *Domecia* dwellings in *M. complanata* were different in shape, usually consisting of folds in the coral's branches (Figure 2). They were most easily observed at the upper margins of the coral's vertical plates (Figure 2d–f) and did not show a collar-like thickening at the margin. Some dwellings occurred at a side of a plate (Figure 2b). One crab was found inside a narrow crevice (Figure 2c). In all cases, the crabs were easy to spot (Figure 2c–f). Interestingly, this association was only observed in exposed reef habitats at Klein Bonaire. It seemed that the crabs altered the shape of some of the *Millepora* corals by giving their branches a more contorted appearance (Figure 2b). Alternatively, the particular form of *M.*

complanata inhabited by *Domecia* may also be a result of environmental factors (surge, currents, and turbulence) to which these organisms are exposed at Klein Bonaire.

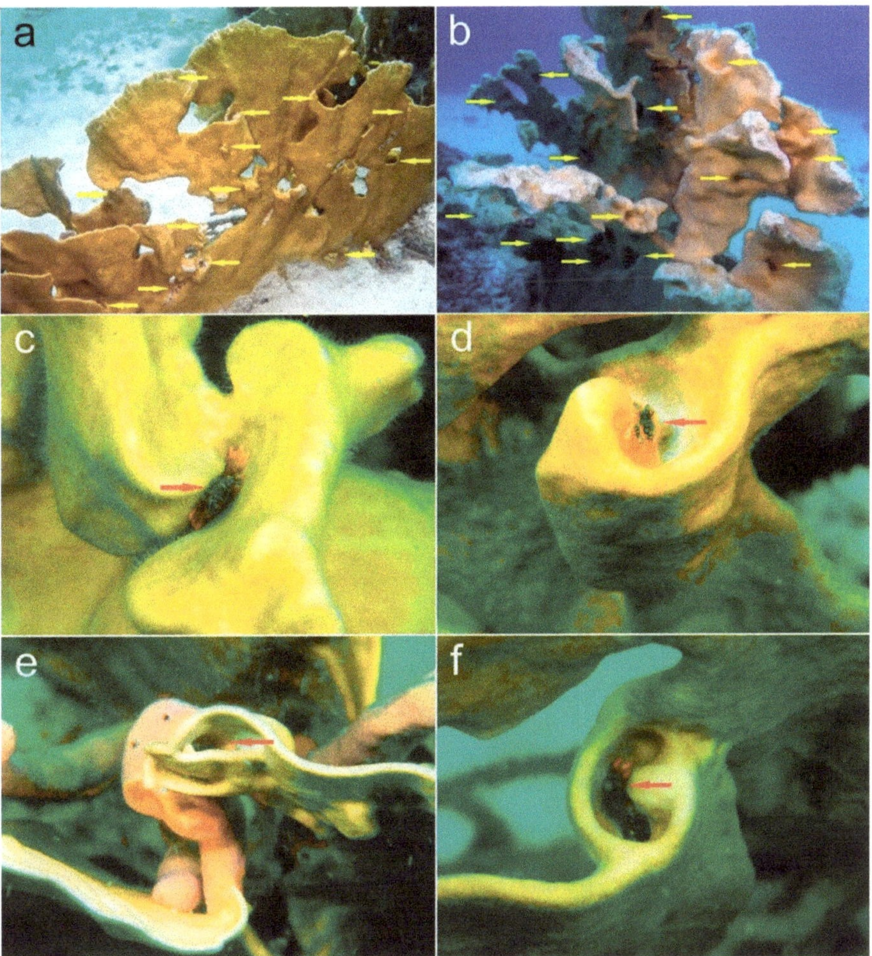

Figure 2. *Millepora complanata* at 2–3 m depth, Klein Bonaire (2019), hosting *Domecia acanthophora*: (**a**,**c**–**f**) Carl's Hill dive site (12°09′53″ N 68°19′23″ W); (**b**) Monte's Divi dive site (12°09′01″ N 68°18′55″ W); (**c**) crab inside narrow crevice; and (**d**–**f**) crabs inside pits consisting of folds at the upper margin of foliaceous coral branches. Yellow arrows: crab dwellings; red arrows: crabs.

The present information may facilitate recognition of endosymbiotic crab fauna in *Acropora* and *Millepora* corals, including fossil ones [4]. Further studies are required to determine whether Atlantic *Domecia* crabs associated with different hosts all belong to *D. acanthophora* or, alternatively, represent distinct but closely related species.

Author Contributions: Conceptualization and supervision, B.W.H.; methodology, illustrations and funding acquisition, B.W.H. and J.E.G.-H.; writing—original draft preparation, B.W.H.; writing—review and editing, B.W.H. and J.E.G.-H. All authors have read and agreed to the published version of the manuscript.

Funding: Fieldwork at Bonaire was supported by the WWF Netherlands Biodiversity Fund and the Treub Maatschappij - Society for the Advancement of Research in the Tropics.

Acknowledgments: We thank staff of CARMABI (Curaçao) for hospitality. The Dive Shop (Curaçao), STINAPA, DCNA and Dive Friends (Bonaire) provided logistic support. Werner de Gier (Naturalis Biodiversity Center) assisted during the field work at Klein Bonaire. We are grateful to three anonymous reviewers for their helpful comments.

Conflicts of Interest: The authors declare no conflict of interest. The funders had no role in the design of the study; in the collection, analyses, or interpretation of data; in the writing of the manuscript, or in the decision to publish the results.

References

1. Castro, P. Brachyuran crabs symbiotic with scleractinian corals: A review of their biology. *Micronesica* **1976**, *12*, 99–110.

2. van der Meij, S.E.T. Host species, range extensions, and an observation of the mating system of Atlantic shallow-water gall crabs (Decapoda: Cryptochiridae). *Bull. Mar. Sci.* **2014**, *90*, 1001–1010. [CrossRef]

3. van der Meij, S.E.T.; Fransen, C.H.J.M.; Pasman, L.R.; Hoeksema, B.W. Phylogenetic ecology of gall crabs (Cryptochiridae) as associates of mushroom corals (Fungiidae). *Ecol. Evol.* **2015**, *5*, 5770–5780. [CrossRef] [PubMed]

4. Klompmaker, A.A.; Portell, R.W.; van der Meij, S.E.T. Trace fossil evidence of coral-inhabiting crabs (Cryptochiridae) and its implications for growth and paleobiogeography. *Sci. Rep.* **2016**, *6*, 23443. [CrossRef] [PubMed]

5. Hoeksema, B.W.; Butôt, R.; García-Hernández, J.E. A new host and range record for the gall crab *Fungicola fagei* as symbiont of the mushroom coral *Lobactis scutaria* at Hawai'i. *Pac. Sci.* **2018**, *72*, 251–261. [CrossRef]

6. Mohammed, T.A.A.; Yassien, M.H. Assemblages of two gall crabs within coral species northern Red Sea, Egypt. *Asian J. Sci. Res.* **2013**, *6*, 98–106. [CrossRef]

7. Terrana, L.; Caulier, G.; Todinanahary, G.; Lepoint, G.; Eeckhaut, I. Characteristics of the infestation of *Seriatopora* corals by the coral gall crab *Hapalocarcinus marsupialis* Stimpson, 1859 on the Great Reef of Toliara, Madagascar. *Symbiosis* **2016**, *69*, 113–122. [CrossRef]

8. Schuhmacher, H. The dwelling cavity of the coral crab *Latopilumnus tubicolus* (Crustacea, Pilumnidae) in *Tubastraea micranthus* (Scleractinia, Dendrophylliidae). *Symbiosis* **1987**, *4*, 289–302.

9. Eldredge, L.G.; Kropp, R.K. *Decapod Crustacean-Induced Skeletal Modification in Acropora*; FAO: Rome, Italy, 1981; Volume 2, pp. 115–119.

10. den Hartog, J.C.; Türkay, M. *Platypodiella georgei* spec. nov. (Brachyura: Xanthidae), a new crab from the island of St. Helena, South Atlantic Ocean, with notes on the genus *Platypodiella* Guinot, 1967. *Zool. Meded.* **1991**, *65*, 209–220.

11. García-Hernández, J.E.; Reimer, J.D.; Hoeksema, B.W. Sponges hosting the Zoantharia-associated crab *Platypodiella spectabilis* at St. Eustatius, Dutch Caribbean. *Coral Reefs* **2016**, *35*, 209. [CrossRef]

12. Patton, W.K. Studies on *Domecia acanthophora*, a commensal crab from Puerto Rico, with particular reference to modifications of the coral host and feeding habits. *Biol. Bull.* **1976**, *132*, 56–67. [CrossRef]

13. Grajal, P.A.; Laughlin, G.R. Decapod crustaceans inhabiting live and dead colonies of three species of *Acropora* in the Roques Archipelago, Venezuela. *Bijdr. Dierk.* **1984**, *54*, 220–230. [CrossRef]

14. Garcia, T.M.; Matthews-Cascon, H.; Franklin-Junior, W. Macrofauna associated with branching fire coral *Millepora alcicornis* (Cnidaria: Hydrozoa). *Thalassas* **2008**, *24*, 11–19.

15. González-Gómez, R.; Briones-Fourzán, P.; Álvarez-Filip, L.; Lozano-Álvarez, E. Diversity and abundance of conspicuous macrocrustaceans on coral reefs differing in level of degradation. *PeerJ* **2018**, *6*, e4922.

Article

Zoantharia (Cnidaria: Hexacorallia) of the Dutch Caribbean and One New Species of *Parazoanthus*

Javier Montenegro [1,2], Bert W. Hoeksema [3,4], Maria E. A. Santos [1], Hiroki Kise [1] and James Davis Reimer [1,2,*]

1   Molecular Invertebrate Systematics and Ecology Laboratory, Graduate School of Engineering and Science, University of the Ryukyus, 1 Senbaru, Nishihara, Okinawa 903-0213, Japan; jmontzalez@gmail.com (J.M.); santos.mariaea@gmail.com (M.E.A.S.); hkm11sea@yahoo.co.jp (H.K.)
2   Tropical Biosphere Research Center, University of the Ryukyus, 1 Senbaru, Nishihara, Okinawa 903-0213, Japan
3   Taxonomy and Systematics Group, Naturalis Biodiversity Center, P.O. Box 9517, 2300 RA Leiden, The Netherlands; bert.hoeksema@naturalis.nl
4   Groningen Institute for Evolutionary Life Sciences, University of Groningen, P.O. Box 11103, 9700 CC Groningen, The Netherlands
*   Correspondence: jreimer@sci.u-ryukyu.ac.jp; Tel.: +81-98-895-8542

http://zoobank.org/urn:lsid:zoobank.org:pub:49D16B6B-BB87-42E8-84F9-F1303EA4EEF2

Received: 6 April 2020; Accepted: 8 May 2020; Published: 12 May 2020

Abstract: Species of the anthozoan order Zoantharia (=Zoanthidea) are common components of subtropical and tropical shallow water coral reefs. Despite a long history of research on their species diversity in the Caribbean, many regions within this sea remain underexamined. One such region is the Dutch Caribbean, including the islands of St. Eustatius, St. Maarten, Saba, Aruba, Bonaire, and Curaçao, as well as the Saba Bank, for which no definitive species list exists. Here, combining examinations of specimens housed in the Naturalis Biodiversity Center collection with new specimens and records from field expeditions, we provide a list of zoantharian species found within the Dutch Caribbean. Our results demonstrate the presence at least 16 described species, including the newly described *Parazoanthus atlanticus*, and the additional potential presence of up to four undescribed species. These records of new and undescribed species demonstrate that although the zoantharian research history of the Caribbean is long, further discoveries remain to be found. In light of biodiversity loss and increasing anthropogenic pressure on declining coral reefs, documenting the diversity of zoantharians and other coral reef species to provide baseline data takes on a new urgency.

Keywords: Anthozoa; coral reefs; records; Macrocnemina; Brachycnemina

1. Introduction

Zoantharians (Anthozoa, Hexacorallia, Zoantharia) are commonly observed rock and reef-dwelling benthic organisms in the Caribbean and other subtropical to tropical regions of the Atlantic. Detailed information about the diversity of the zoantharian fauna at species level has been reported from various parts of the Atlantic Ocean and adjacent seas, although some records may not be complete or taxonomically not up to date anymore. For example, 13 species have been recorded from the coastline of Brazil [1], 11 from of the Gulf of Mexico [2], 10 from the Bahamas and Florida [3], 8 from the Cape Verde islands in the eastern Atlantic [4] and Saint Helena Island [5], 7 from Bermuda [6], 7 from the Mediterranean [7], 5 in Canary Islands [8], and 4 from Ascension Island in the central Atlantic [9]. In addition, five deep-water species from the Azores (eastern Atlantic) were described [10,11]. While some recent progress has been made in assessing oceanic basin-wide diversity patterns [5], questions remain on the true numbers of species present in each region due to a general lack of field data and ambiguous or brief original species descriptions. Thus, detailed data with accurate species identifications, the basis of large-scale biodiversity analyses, are still needed from most regions of the Atlantic.

Information about the zoantharian fauna of the Dutch Caribbean is generally poor despite a long history of biodiversity research [12]. The Dutch Caribbean consists of the islands Aruba, Bonaire, and Curaçao off Venezuela in the southern Caribbean region, and the islands Saba, St. Eustatius, and St. Maarten, together with the submerged platform Saba Bank, in the eastern Caribbean region. Most zoantharian records from these regions concern shallow water specimens and are usually not identified to more specific than the genus level. For instance, Van der Horst [13] refers to "Zoanthacea" as social sea anemones covering rocks that partly reach above the seawater level in Caracas Bay in Curaçao Island. Wagenaar Hummelinck [14,15] reported *Zoanthus* in shallow waters of Aruba and Curaçao. Van den Hoek et al. [16] mentioned that the genera *Palythoa* and *Zoanthus* were common in Curaçao in waters less than ca. 1.2 m deep. Wanders [17] and Nagelkerken and Nagelkerken [18] only reported on the presence of zoantharians in benthic communities of the shallow reef zones here.

However, there is some detailed, species-level information on the zoantharian fauna in the Dutch Caribbean. For example, Bak [19] mentioned *Palythoa mammillosa* as living in very shallow water of the shore zone in front of cliffs of Bonaire and Curaçao. Van Duyl [20] refers to zoantharians in general and to *Palythoa caribaeorum* when discussing the shallow water ecosystems of these same two islands. More recently, Reimer [21] distinguished 14 species in a preliminary report from a marine biodiversity expedition in St. Eustatius, and Garcia-Hernandez et al. [22] reported associations between *Palythoa caribaeorum* (Duchassaing de Fonbressin & Michelotti, 1860) [23] and *Umimayanthus parasiticus* (Duchassaing de Fonbressin & Michelotti, 1860) [23] and a crab species at St. Eustatius. Finally, Reimer et al. [24] have provided a detailed list of the zoantharian species of the west coast of Curaçao on coral reefs to 30 m depth.

The present report aims to give an update on the zoantharian fauna throughout the Dutch Caribbean as a result of recent fieldwork (2014–2019) by the authors in Bonaire, Curaçao, and St Eustatius, during which Zoantharia specimens were photographed in situ and collected. In addition, previously collected specimens deposited in the zoological collections of Naturalis Biodiversity Center were newly studied and identified to genus or species-level. In this manner, we have been able to cover a wide range of species across the Dutch Caribbean. Since most previous publications dealing with the zoantharians in the Dutch Caribbean were not performed by zoantharian specialists, it is hoped that the present work will serve as a basis for others performing research on zoantharians in the Atlantic and particularly in the Caribbean.

2. Materials and Methods

2.1. Specimens Analyzed

Our examinations included specimens from the Coelenterata and Porifera collections (RMNH and ZMA) at Naturalis Biodiversity Center in Leiden, Netherlands, and from the Molecular Invertebrate Systematics and Ecology Laboratory (MISE) collection at the University of the Ryukyus in Okinawa, Japan. Specimens and surveys included the island states of Aruba, Curaçao, Sint Maarten (both Netherlands and French territories), and the Caribbean Netherlands including the islands of Bonaire, Sint Eustatius, and Saba, as well as the submerged Saba Bank. In total, 479 zoantharian specimens were analyzed in this study (Table S1), including 181 specimens belonging to the Naturalis collections (60 from the Porifera collection, 121 from the Coelenterata collection including 6 specimens were collected in Curaçao by the second author from the shallow and deep sea in 2014), and 298 to the MISE collection. Among the specimens from the MISE collection, 173 were collected in the Sint Eustatius survey of 2015 [21], 86 were collected in the Curaçao survey of 2017 [24], and 39 were collected in the Bonaire survey of 2019.

2.2. Specimen Identification

Most specimens in this study were identified by the first or last author between 2012 and 2019. For identification, we focused on external morphological characteristics that are utilizable in the field (e.g., general colony morphology, polyp sizes, tentacle numbers [25,26]). All measurement units were converted to the standard international metric system. The large majority of newly collected specimens in this study were identified via simple morphological and ecological analyses. We additionally conducted molecular phylogenetic analyses for one species that we formally describe in this work; these methods are given below.

Previously collected and identified specimens from earlier field work were also re-identified as much as possible by the first or last author (n = 173), although some of these earlier specimens (particularly type specimens) retain their original identification with no further amendment (n = 2). Such 'earlier' identifications, however, may be synonymous with other species [5,27,28], and have not been counted in species totals following the methodology in Santos et al. [5], and are instead listed within species groups (as "b", "c", etc, see Table 1).

Table 1. Depth distributions of Zoantharia species in the Dutch Caribbean. The Southern Caribbean region includes the islands of Aruba, Bonaire, and Curacao, while the Eastern Caribbean covers Saba, Saba Bank, Saint Eustatius, and Saint Maarten. We divided records into historic (earlier than 2000) and recent surveys (2001 and later). References for each species' depths are included in each species' section of the Results.

Index	Species	Range (m)	Southern Caribbean (m)		Eastern Caribbean (m)	
			Historic	Recent	Historic	Recent
1	Parazoanthidae sp. 1	140–248	x	140–248	x	x
2	*Antipathozoanthus* aff. *macaronesicus*	10	10	x	x	x
3	*Bergia catenularis*	6–50	6–36	10–38	20–50	12–26
4	*Bergia* cf. *cutressi* *	0–8	0–8	x	x	x
5	*Bergia puertoricense* *	10–55	10–55	10–38	x	13–37
6	*Parazoanthus swiftii*	10–40	10	19–40	unknown	14–29
7	Parazoanthidae? sp.	unknown	unknown	x	x	x
8	*Parazoanthus atlanticus*	10–34	x	10–34	x	x
9	*Umimayanthus parasiticus* *	1–44	1–40	15–38	16–44	3–34
10	*Umimayanthus* sp.	37	x	37	x	x
11	*Epizoanthus* sp.	980	x	x	980	x
12	*Hydrozoanthus antumbrosus* *	11–30	x	30	x	11–19
13	*Hydrozoanthus tunicans* *	2–30	2–4	30	x	14–19
14a	*Palythoa caribaeorum*	0–35	0–2	0–14	0–35	2–29
14b	Palythoa caracasiana *[†]	unknown	unknown	x	x	x
14c	*Palythoa horstii* *[†]	unknown	unknown	x	x	x
14d	*Palythoa mammilosa* *[†]	2	unknown	x	2	x
15	*Palythoa grandiflora*	1–6	intertidal	x	1–6	x
16	*Palythoa grandis*	11–64	18–64	11–12	x	13–18
17	*Palythoa variabilis*	0–37	0–24	37	intertidal	3
18	*Palythoa* sp.	intertidal	intertidal	x	x	x
19	*Zoanthus pulchellus*	0–24	0–24	1–11	0–20	15–16
20	*Zoanthus* aff. *pulchellus*	1	x	1	x	x
21	*Zoanthus sociatus*	0–24	0–3	intertidal	0–6	3–24
22	*Zoanthus solanderi*	0–21	intertidal	12–16	0–15	3–21
23	*Zoanthus* sp.	intertidal	intertidal	x	intertidal	x
24	*Isaurus tuberculatus*	0–15	intertidal	x	2	15
Diversity recorded		24	18	16	14	15
			22		17	

* Species endemic to the Caribbean Sea. x = not reported. [†] Species most likely synonym of *P. caribaeorum*, and therefore not included in the species counts.

Zoantharian specimens belonging to the Naturalis collection were identified or re-identified using the gross external morphology of preserved specimens, and biological interactions/associations when applicable (e.g., [24,29]). All MISE specimens from Sint Eustatius, Bonaire, and Curaçao were identified using in situ or in vivo images.

For each species listed in the Results, we have also included a description following as close as possible the original description, with some small amendments to reflect additional information acquired after the first formal description. These descriptions should not be interpreted as formal descriptions, to the exception of the one new species we describe here, but as information provided to workers to aid in field identification and to make the original descriptions accessible. A list of specimens, their collection information, and Naturalis or MISE registration numbers are given within each species section.

2.3. Cnidae Analyses

Analyses were conducted using undischarged nematocysts from tentacles, column, actinopharynx, and mesenteries filaments of holotype polyps ($n = 2$; specimen NSMT-Co 1706) under a Nikon Eclipse80i stereomicroscope (Nikon, Tokyo, Japan). Cnidae sizes were measured using ImageJ ver. 1.45s [30]. Cnidae classification generally followed England [31] and Ryland and Lancaster [32], while basitrichs and microbasic b-mastigophores were considered as the same type of nematocyst based on studies by Schmidt [33], Hidaka et al. [34], and Hidaka [35] and therefore these two types were pooled together.

2.4. DNA Extraction, Polymerase Chain Reaction (PCR) Amplification, and Sequencing

Total DNA was extracted using the Qiagen DNeasy Blood & Tissue Kit following the manufacturer's instructions for specimens NSMT-Co 1706, NSMT-Co 1707, MISE JDR170613-10-60, MISE JDR170613-10-61, MISE JDR170616-13-76, RMNH.COEL.42433, MISE JDR170609-2-6, MISE JDR170610-4-32, and MISE JDR170619-20-94. PCR amplification was performed for partial sequences of cytochrome oxidase subunit I (COI-mtDNA) following Folmer et al. [36], mitochondrial 16S ribosomal DNA (16S-rDNA) following Sinniger et al. [37], and the nuclear internal transcribed spacer region of ribosomal DNA (ITS-rDNA) following Reimer et al. [38] using standard Taq polymerase in ReadyMix solution (Qiagen, Tokyo, Japan). Successful amplifications were confirmed by 2% agarose gel electrophoresis, cleaned by shrimp alkaline phosphatase (SAP), and sent for external sequencing in both directions to Fasmac, Kanagawa, Japan.

2.5. Phylogenetic Analyses and Species Delimitations

The nucleotide sequences were initially aligned using Geneious v10.2.3 [39] and the plugin MAFFT [40] with the algorithm L-INS-i, thereafter the sequences were manually curated and trimmed. Trimmed alignments were subsequently realigned using the plugin MUSCLE [41] in Geneious v10.2.3 with default settings and aligned with previously reported sequences from family Parazoanthidae found in GenBank (Table 2). The resulting alignments were 446 sites of 30 sequences for COI-mtDNA, 576 sites for 51 sequences for 16S-rDNA, and 814 sites for 44 sequences for ITS-rDNA. These three alignments were then used to construct a concatenated alignment; missing data and gabs were replaced with "Ns". The final concatenated alignment consisted of 1836 sites and 57 sequences (Table 2). All alignments are available from the first and senior authors, and at treebase.org (ID: 26174).

Table 2. List of all Zoantharia sequences used in phylogenetic analyses, and their respective sequences GenBank ID number. NA = not available.

Species/Specimens	COI-mtDNA	16S r-DNA	ITS r-DNA
Antipathozoanthus macaronesicus	NA	HM130467	EU591552
Bergia catenularis	NA	EU828757	EU418289
Bergia catenularis (TOB37)	NA	NA	EU418292
Bergia cutressi (1)	NA	EU828759	EU418264
Bergia cutressi (2)	NA	NA	EU418267
Bergia puertoricense (1)	AB247351	AY995933	EU591584
Bergia puertoricense (2)	NA	EU828758	EU418312
Bergia sp. Senegal	EF672656	EF687820	EU591582
Bergia sp. 5 Sulawesi	EU591627	AY995934	NA
Bullagummizoanthus emilyacadiaarum	NA	KC218434	NA
Corallizoanthus tsukaharai	NA	EU035625	EU035621
Epizoanthus arenaceus	AB247348	AY995926	EU591538
Hurlizoanthus parrishi	NA	KC218433	NA
Isozoanthus giganteus	NA	GQ464867	GQ464896
Kauluzoanthus kerbyi (SH12)	NA	KC218435	NA
Kulamanamana haumeaae (SH2)	NA	KC218431	NA
Mesozoanthus fossii	NA	EF687822	EU591545
Parazoanthid sp. 02_27	NA	EU333760	EU333810
Parazoanthid sp. 3 Madagascar	EF672664	EF687825	EU591576
Parazoanthid sp. Tasmania	EU591620	EU591610	NA
Parazoanthid sp. 3 Sulawesi	AB247354	AY995937	EU591575
Parazoanthus aff. *juanfernandezii* (CA128)	NA	GQ464849	GQ464878
Parazoanthus aff. *swiftii* (PER241)	NA	GQ464853	GQ464882
Parazoanthus aff. *swiftii* (PER249)	NA	GQ464854	GQ464883
Parazoanthus anguicomus (1)	EF672660	EF687827	EU591574
Parazoanthus anguicomus (2)	NA	GQ464851	GQ464880
Parazoanthus axinellae (1)	AB247355	AF398921	NA
Parazoanthus axinellae (2)	EF672659	NA	EU591571
Parazoanthus capensis (SA262)	NA	GQ464852	GQ464881
Parazoanthus darwini (1)	NA	EU333748	EU333802
Parazoanthus darwini (2)	NA	EU333751	NA
Parazoanthus elongatus (Chile)	EF672661	EF687829	EU591565
Parazoanthus elongatus (NZ)	EF672662	EF687828	EU591564
Parazoanthus sp. 1401	NA	HM130478	NA
Parazoanthus sp. 269	NA	HM130468	NA
Parazoanthus sp. 'hertwigi'	KC218397	NA	NA
Parazoanthus swiftii (1)	AB247350	AY995936	GQ848258
Parazoanthus swiftii (2)	KJ794176	EU828755	EU418332
Savalia savaglia	NA	HQ110948	EU346888
Umimayanthus chanpuru (16J)	KR092609	KR092469	KR092678
Umimayanthus chanpuru (33J)	KR092594	KR092504	KR092680
Umimayanthus miyabi (179TF)	KR092570	KR092453	KR092645
Umimayanthus miyabi (70JR)	KR092573	KR092454	KR092646
Umimayanthus nakama (363JR)	KR092577	KR092458	KR092644
Umimayanthus nakama (3J)	KR092579	KR092457	KR092643
Umimayanthus parasiticus (1)	EF672663	AY995938	GQ848263
Umimayanthus parasiticus (2)	NA	EU828756	EU418306
Zibrowius ammophilus (SH15)	NA	KC218439	NA
Parazoanthus atlanticus sp. n. (RMNH.COEL.42433)	NA	NA	MT103525
Parazoanthus swiftii (MISE JDR170609-2-6)	MT102228	MT103533	MT103530
Parazoanthus swiftii (MISE JDR170610-4-32)	MT102229	MT103534	MT103531
Parazoanthus atlanticus sp. n. (MISE JDR170613-10-60)	MT102223	MT103538	MT103528
Parazoanthus atlanticus sp. n. (MISE JDR170613-10-61)	MT102222	MT103539	MT103527
Parazoanthus atlanticus sp. n. (NSMT-Co 1706)	MT102224	MT103537	MT103526
Parazoanthus atlanticus sp. n. (NSMT-Co 1707)	MT102225	MT103536	MT103524
Parazoanthus atlanticus sp. n. (MISE JDR170616-13-76)	MT102226	MT103535	MT103529
Umimayanthus sp. (MISE JDR170619-20-94)	MT102227	NA	NA

Phylogenetic analyses were performed on the concatenated aligned dataset using maximum-likelihood (ML) and Bayesian posterior probability (BPP). TOPALi v2.5 [42] was used to select the best fitting model for each COI-mtDNA, 16S-rDNA, and ITS-rDNA regions, independently for ML and BPP analyses. For ML analyses, the best-fitting models were K80+G (010010), TrNef+G (010020), and HKY+G (010010); and for BPP K80+G (010010), K80+G (010010), and HKY+G (010010), respectively, for COI-mtDNA, 16S-rDNA, and ITS-rDNA regions. Independent phylogenetic analyses were performed for each region and for the concatenation in RAxML v8.2.11 [43] for ML, and Mr. Bayes v3.2.6 [44] for BPP. RAxML was configured to use the substitution model GTR+G with the "-f a –x 1" algorithm, 1000 bootstrap replicates, 1 parsimony random seed, and *Epizoanthus arenaceous* was specified as out-group. MrBayes was configured following the models and parameters as indicated by TOPALi, 4 MCMC heated chains were run for 10,000,000 generations with the temperature for the heated chain set to 0.2. Chains were sampled every 200 generations. Burn-in was set to 3,500,000 generations (35%), at which point the average standard deviation of split frequency (ASDOSF) values were <0.01.

For specimens with molecular sequences available, species identifications were determined using a combination of molecular and morphological data. Species were delimited according to monophyletic clades of our generated concatenated phylogenetic tree (genealogical species concept; [45,46]), and their validity was evaluated using available morphological characters.

3. Results

3.1. Diversity in the Dutch Caribbean

Overall, 126 unique locations were examined across the study area (Figure 1a,b; Table S2), although the specific collection site information for 20 specimens were not available. Four localities were investigated in Aruba, 30 in Bonaire, 44 in Curaçao, 6 in Saba, 11 in Saba Bank, 26 in Sint Eustatius, and 5 in Sint Maarten, 3 of which were located in French territory.

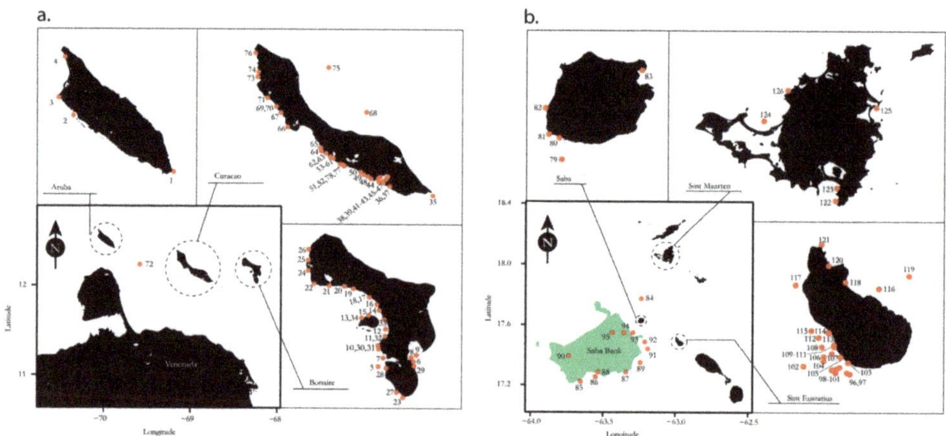

Figure 1. Map of the Dutch Caribbean islands. (**a**) Southern Caribbean region with Aruba, Bonaire, and Curaçao islands. (**b**) Eastern Caribbean region including Saba, Saba Bank, Sint Maarten, and Sint Eustatius. Red points indicate the approximate position of all localities included in this study. For detailed information on localities see Table S2 [47,48].

Most Zoantharia specimens were easily identifiable to species level, to the exception of 3 specimens that were identified as "confers with" (cf.), 4 as "affinity" (aff.), 14 to genera, and 5 only to supra-generic levels. In total, all specimens studied represented 9 genera and 17 described species.

Four potentially undescribed species were found in this survey; one undescribed species belonging to genus *Umimayanthus* (sample ID: MISE JDR170619-20-94, MISE JDR191026-1-1) and one to *Epizoanthus* (RMNH.COEL.40667), while specimens RMNH.COEL.42429, RMNH.COEL.42430; RMNH.POR. 9219, 9234, and 9251 possibly belong to one or two species in a potentially undescribed Parazoanthidae genus from the deep sea around Curaçao Island. Additionally, in this study, we formally describe one species belonging to genus *Parazoanthus* from six specimens (RMNH.COEL.42433; NSMT-Co 1706 and NSMT-Co 1707; JDR170613-10-60 to 61; JDR170616-13-76) that superficially resembled *P. swiftii* (Duchassaing de Fonbressin & Michelotti, 1860) [23] but was shown to have clear differences in habitat, polyp size, colony arrangement, and in molecular data from *P. swiftii* and other *Parazoanthus* species.

Specimens are listed by species with information in the following order: Specimen number, latitude and longitude, location, depth, date, collectors.

3.2. Specimens and Species

Order Zoantharia Rafinesque, 1815 [49]
Suborder Macrocnemina Haddon & Shackleton, 1891 [50]
Family Parazoanthidae Delage & Hérouard, 1901 [51]
Figures 2–26

3.2.1. Parazoanthidae sp. (Figure 2)

Specimens examined (*n* = 5). **Bonaire**. RMNH.POR.9234 (12°04′48″ N, 68°17′38″ W [point 5], Curasub, Cargill Pier, 223 m depth, 1.vi.2013, coll. L.E. Becking & E.H.W.G. Meesters); RMNH.POR.9251 (similar but from 248 m depth); RMNH.POR.9219 (12°08′49″ N, 68°16′56″ W [point 12], Kralendijk Pier, 140 m depth, 30.v.2013, coll. L.E. Becking & E.H.W.G. Meesters). **Curaçao** RMNH.COEL.42429 (12°14′01″ N, 68°53′32″ W [point 68], Curasub, Playa Porto, Curaçao, 61–243 m depth, 21.iii.2014, coll. BWH); RMNH.COEL.42430 (12°05′04″ N, 68°53′54″ W [point 49], Curasub, Substation Curaçao, ca. 200 m depth, 31.iii.2014, coll. BWH).

Photographic records (*n* = 5). In situ: Specimens RMNH.COEL.42429, RMNH.COEL.42430. Preserved: Specimens RMNH.POR.9219, 9234, 9251.

Remarks: All specimens were collected from the deep sea (depths 140–248 m), as symbiont of sponges. The examined specimens have cyclically transitional or cteniform marginal musculature, and these marginal musculature forms have been reported from the family Parazoanthidae [52,53]. Within this family, several *Isozoanthus* species are known to have association with stalked hexactinellid sponges within subclass Amphidiscophora Schulze, 1886 [54,55]. However, the examined specimens are associated with hexactinellid sponges within subclass Hexasterophora Schulze, 1886. Therefore, we consider that these specimens belong to an undescribed genus possibly containing more than one species. For now, we have listed specimens as one taxon in this work. Photographs of this species were published earlier as "zoanthids" living in association with the sponges *Cyrtaulon sigsbeei* (Schmidt, 1880) and *Verrucocoeloidea liberatorii* Reiswig & Dohrmann, 2014 (see [56] (Figure 4a,b); [57] (Figure 22f,g)), and were also recorded from Bonaire and Curaçao.

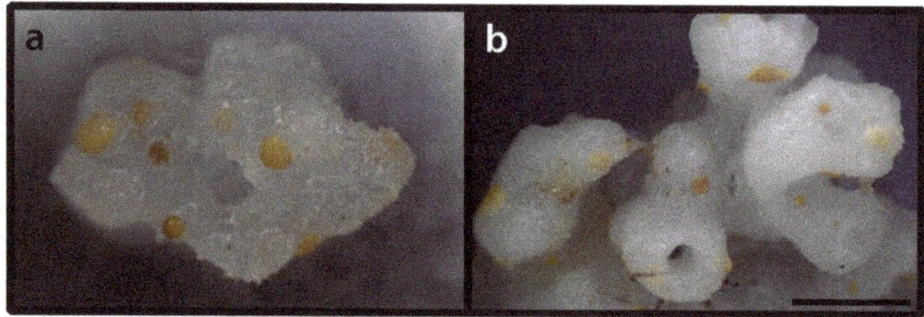

Figure 2. Preserved specimens of Parazoanthidae sp. from Naturalis collected from Bonaire; (**a**) specimen RMNH.POR.9219 from Kralendijk Pier [point 12], depth = 140 m, and (**b**) RMNH.POR.9234 from Cargill Pier [point 5], depth = 223 m. Scale bar in (**b**) = approximately 1 cm.

Genus *Antipathozoanthus* Sinniger, Reimer & Pawlowski, 2010 [58]

3.2.2. *Antipathozoanthus* aff. *macaronesicus* Ocaña & Brito, 2003 [59] (Figure 3)

Specimens examined (*n* = 3). **Curaçao.** RMNH.COEL.40331, 40332, 40763 (12°04′30″ N, 68°51′51″ W [point 45], Caracas Baai, Buoy 9, >10 m depth, 9.ii.1955, coll. J.S. Zaneveld & P. Wagenaar Hummelinck).

Photographic records (*n* = 3). Preserved: Specimens RMNH.COEL.40331, 40332, 40763.

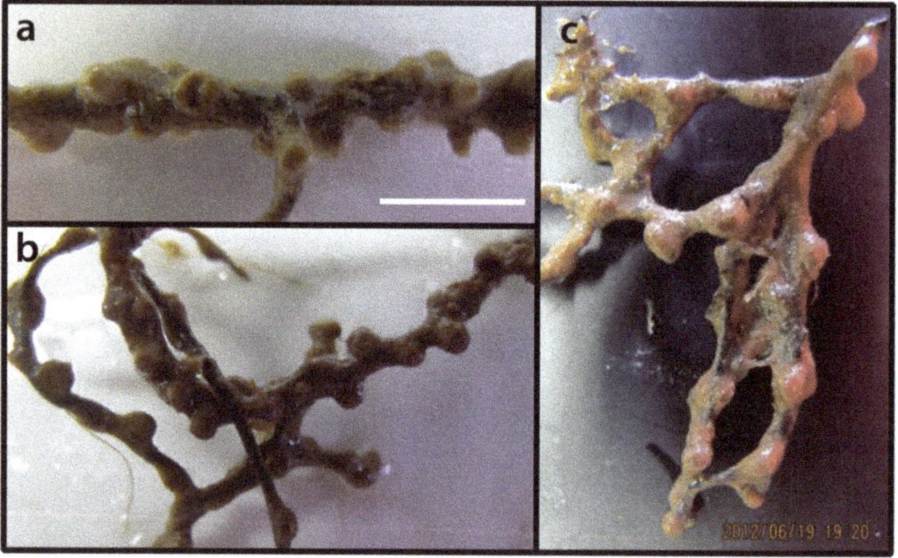

Figure 3. Preserved specimens of *Anthipathozoanthus* aff. *macaronesicus* from Naturalis collected from Buoy 9, Caracas Baai [point 45], Curaçao, all depth >10 m; (**a**) specimen RMNH.COEL.40331, (**b**) RMNH.COEL.40332, and c) RMNH.COEL.40763. Scale bar in (**a**) = approximately 1 cm.

Description of *A. macaronesicus* adapted from Ocaña & Brito [59]. Colonies present several forms of growth and a variable external appearance. When growing freely, the colony develops its own skeleton and generates branches in a single or multiple direction; although it may also present a poorly developed ribbon-like skeleton. The colony can also grow over anthipatharians and reach up to one meter in height. In preserved material, polyp dimensions are variable from 0.2 to 1 cm in height and 0.2 to 0.5 in diameter; alive, the polyp sizes increase considerably to 2 to 3 cm. Tentacles are pointed, 42 in number, and, more or less, arranged in two entacmeic cycles. Polyps are overhanging from coenenchyme, but embedded polyps are also present. The colony can present several colors ranging from yellow to orange, in tentacles and column; colonies growing in antipatharians typically present brown color.

Remarks: All specimens were collected from colonies growing on antipatharians. Specimen RMNH.COEL.40332 was growing on an unidentified antipatharian. RMNH.COEL.40763 was growing on *Antipathes gracilis* Gray, 1860. No *Antipathozoanthus* were found during our recent surveys of Curaçao, and all known specimens were collected in 1955. Currently, only one species, *A. macaronesicus* (Ocaña & Brito, 2003) [59], is known for the genus *Antipathozoanthus* from the Atlantic. However, four species of the genus have been described from the Indian and Pacific Oceans [60,61]. Thus, given the distance from confirmed records of *A. macaronesicus* in the East Atlantic, we here identify all specimens as *Antipathozoanthus* aff. *macaronesicus*.

Genus *Bergia* Duchassaing de Fonbressin & Michelotti, 1860 [23]

3.2.3. *Bergia catenularis* Duchassaing de Fonbressin & Michelotti, 1860 [23] (Figure 4)

Specimens examined (*n* = 26). **Bonaire**. MISE JGH191024-2-1 (12°7′53.28″ N, 68°16′59.76″ W [point 32], Corporal Meiss, 27 m depth, 24.x.2019, coll. JGH); MISE JDR191025-1-1 (12°12′1.74″ N, 68°18′30.72″ W [point 18], Oil Slick, 17 m depth, 25.x.2019, coll. JDR); MISE JDR191025-2-2 (12°2′8.22″ N, 68°15′43.32″ W [point 27], Sweet Dreams, 30 m depth, 25.x.2019, coll. JDR); MISE JDR191103-2-6 (12°1′36.3″ N, 68°15′4.74″ W [point 23], Red Slave, 20 m depth, 3.xi.2019, coll. JDR); MISE JDR191103-2-7 (12°1′36.3″ N, 68°15′4.74″ W [point 23], Red Slave, 20 m depth, 3.xi.2019, coll. JDR). **Curaçao**. MISE JDR170610-3-26 (12°08′21″ N, 68°59′53″ W [point 64], Snake Bay, 31 m depth, 10.vi.2017, coll. JDR); MISE JDR170610-4-31 (12°08′53″ N, 69°00′00″ W [point 65], Sint Michiel's Bay, 20 m depth, 10.vi.2017, coll. JDR); MISE JDR170610-4-33 (12°08′53″ N, 69°00′00″ W [point 65], Sint Michiel's Bay, 10 m depth, 10.vi.2017, coll. JDR); MISE JDR170621-night-102 (12°07′20″ N, 68°58′08″ W [point 54], Carmabi, House Reef, Curaçao, unknown depth, 21.vi.2017, coll. J.E. Garcia-Hernandez); MISE NA (12°19′45″ N, 69°09′05″ W [point 74], Playa Jeremi, 12–38 m depth, 20.vi.2017, coll. JDR); ZMA.POR.14242 (12°08′21″ N, 68°59′53″ W [point 64], Snake Bay, 36 m depth, 18.iv.1989, coll. M.J. de Kluijver); ZMA.POR.14344 (12°07′32″ N, 68°58′27″ W [point 59], north of Piscadera Bay, Buoy 1, 35 m depth, 15.v.1998, coll. R. Gomez); ZMA.POR.15665 (12°08′01″ N, 68°59′07″ W [point 62], Blue Bay, 35 m depth, 25.ii.1989, coll. R.W.M. van Soest); ZMA.POR.19055 (12°06′33″ N, 68°57′15″ W [point 52], Santa Marta, Water Factory, unknown depth, 2005, coll. N. van der Hal); ZMA.POR.4626 (12°07′30″ N, 68°58′23″ W [point 58], north of Piscadera Bay, Buoy 0, 6–12 m depth, 19.xii.1980, coll. R.W.M. van Soest). **Saba Bank**. ZMA.POR.5143 (17°14′00″ N, 63°34′00″ W [point 86], Sta. LUY-101, south-slope, 20–50 m depth, 24.x.1972, coll. Luymes Exp.). **Sint Eustatius**. MISE JDR150610-6, JDR150610-7 (17°27′44.2″ N, 62°58′46.7″ W [point 97], Sta. EUX007, 21 m depth, 10.vi.2015, coll. JDR; MISE JDR150610-12 (17°27′53.9″ N, 62°59′00.7″ W [point 101], Sta. EUX008, 17 m depth, 10.vi.2015, coll. JDR); MISE JDR150611-33, JDR150611-34 (17°28′19.2″ N, 62°59′15.6″ W [point 107], Sta. EUX010, 12 m depth, 11.vi.2015, coll. JDR); MISE JDR150612-81 (17°30′57.4″ N, 62°59′21.6″ W [point 120], Sta. EUX011, 16 m depth, 12.vi.2015, coll. JDR); MISE JDR150614-127 (17°27′50.9″ N, 62°59′06.8″ W [point 100], Sta. EUX015, 16 m depth, 14.vi.2015, coll. JDR); MISE JDR150616-147 (17°28′13.6″ N, 62°59′30.2″ W [point 106], Sta. EUX019, 18 m depth, 16.vi.2015, coll. JDR); MISE JDR150618-157 (17°27′56.6″ N, 63°00′07.2″ W [point 102], Sta. EUX022, 26 m depth, 18.vi.2015, coll. JDR); MISE JDR150619-166 (17°31′35.7″ N, 62°59′35.3″ W [point 121], Sta EUX024, 25 m depth, 19.vi.2015, coll. JDR).

Photographic records (*n* = 24). In situ: Specimens MISE JDR150610-6, JDR150610-7, JDR150610-12, JDR170610-3-26, JDR170610-4-31, JDR170610-4-33, JDR150611-33, JDR150611-34, JDR150612-81, JDR150614-127, JDR150616-147, JDR150618-157, JDR150619-166, JGH191024-2-1, JDR191025-1-1,

JDR191025-2-2, JDR191103-2-6, JDR191103-2-7. Preserved: Specimens ZMA.POR.4626, 5143, 14242, 14344, 15665, 19055.

Description as in Duchassaing de Fonbressin & Michelotti [23], West [26], Swain [62]. The colonies present very short polyps forming a chain-like incrustation on the surface of the sponges, with polyps arising from one another by stolons (propagules), not from a common membrane. This species is characterized by having a commensalistic and cateniform habit of colonization (p. 54 translated by Duerden [63] from Duchassaing de Fonbressin & Michelotti [23], also [29]). Polyps and coenenchyme present a golden–brown color, 10 capitular ridges, 20 tentacles with a maximum length of 1 mm, and the same number of mesenteries, the length and diameter of extended polyps is rarely more than 1 mm, and the symbiont sponges belonging to the order Halosclerida [26,62].

Recent and other previous records: Antilles [23], Bahamas [64], Barbados [65,66], Brazil [66], Colombia [67], Curaçao [24], Dominica [66], Jamaica [68], Panama [66], Puerto Rico [26], Tobago [66], USA (Gulf of Mexico, Navassa Island) [66,69], and Venezuela [70].

Remarks: This species, originally placed with the genus *Bergia*, was placed into *Parazoanthus* Haddon & Shackleton, 1891 [50] by Duerden [63], and thus appears in most literature as *P. catenularis*, until the resurrection of *Bergia* based on molecular data by Montenegro et al. [71]. *B. catenularis* was observed associated with *Petrosia* (*Petrosia*) aff. *weinbergi* Van Soest, 1980 in Curaçao [24] and Saba Bank; and with *Xestospongia muta* (Schmidt, 1870) in Sint Eustatius.

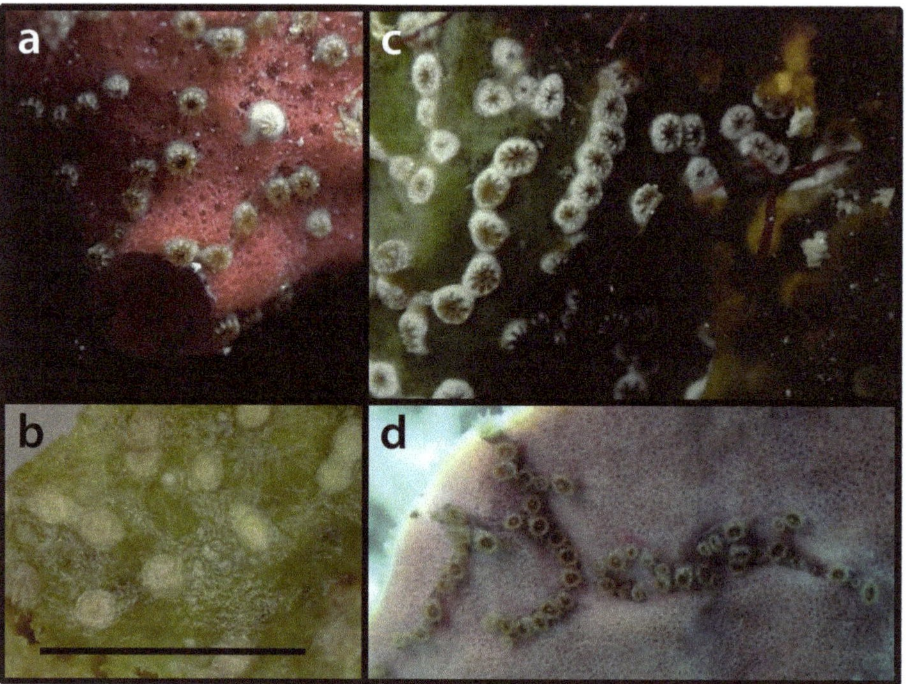

Figure 4. Specimens of *Bergia catenularis*. (**a**) specimen MISE JDR170610-4-31 in situ from Sint Michiel's Bay [point 65], Curaçao, depth = 20 m, (**b**) preserved ZMA.POR.19055 collected from Santa Marta, Water Factory, Curaçao, depth = unknown, (**c**) MISE JDR170610-4-33 in situ from Sint Michiel's Bay [point 65], Curaçao, depth = 10 m, and (**d**) MISE JDR150611-34 from Sta. EUX010 [point 107], Sint Eustatius, depth = 12 m. Scale bar in (**b**) = approximately 1 cm.

3.2.4. *Bergia* cf. *cutressi* (West, 1979) [26] (Figure 5)

Specimens examined (*n* = 3). **Bonaire**. MISE JDR191029-1-2 (12°13′24.42″ N, 68°24′13.38″ W [point 22], Taylor Made, 29 m depth, 29.x.2019, coll. JDR); MISE JDR191107-1-1, (12°6′37.74″ N, 68°17′35.16″ W [point 30], The Lake, 13 m depth, 7.xi.2019, coll. JDR). **Sint Maarten.** RMNH.COEL.40278 (18°00′31″ N, 63°02′46″ W [point 122], Great Bay near Pointe Blanche, Sta. LUY-120, 0–8 m depth, 27.ix.1972, coll. JCH).

Photographic records (*n* = 3). In situ: MISE JDR191029-1-2, JDR191107-1-1. Preserved: Specimen RMNH.COEL.40278.

Description as in West [26]: Colonies embedded in *Xestospongia* sp. sponge, with only scapus of polyps projecting from the surface. The polyp dimensions are 1 mm in high and 0.3 mm in diameter. The polyps are connected beneath the sponge surface by a coenenchyme; scapulus thin-walled and clean, capitular ridges are 12 in number; tentacles and mesenteries are 12 in number. Coenenchyme, column, and tentacles are the color yellow.

Recent and other previous records: Barbados [65,66], Colombia [67], Dominica [66], Puerto Rico [26], Tobago, and USA (Navassa) [66].

Remarks: Similar to *B. catenularis*, this species was originally described as part of *Epizoanthus* Gray, 1867 [72] by West [26], later placed into *Parazoanthus* by Swain et al. [53], until the resurrection of *Bergia* by Montenegro et al. [71], where molecular evidence placed this species within genus *Bergia*. Thus, this species still appears in most literature as *E. cutressi*. The associated host sponge of specimen RMNH.COEL.40278 was not identified.

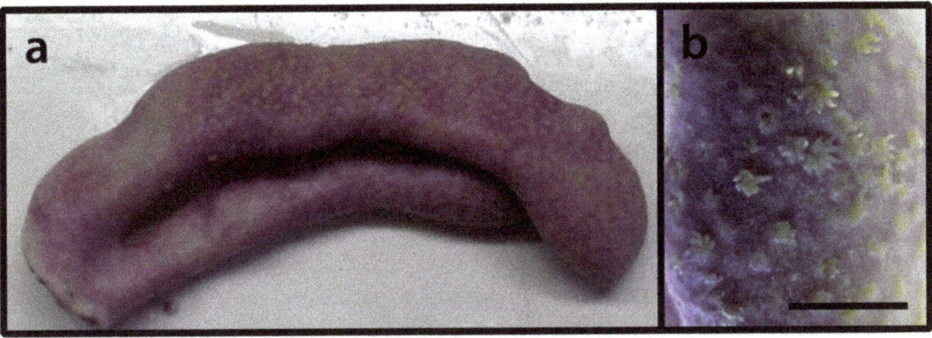

Figure 5. (**a**) Preserved *Bergia* cf. *cutressi* specimen RMNH.COEL.40278 from Great Bay near Pointe Blanche, Sta. LUY-120 [point 122], Sint Maarten, depth = 0–8 m, (**b**) and close-up of same specimen. Scale bar in (**b**) = approximately 1 cm.

3.2.5. *Bergia puertoricense* (West, 1979) [26] (Figure 6)

Specimens examined (*n* = 29). **Bonaire**. MISE JGH191106-2-1 (12°11′17.1″ N, 68°17′47.7″ W [point 16], Andrea I, 17 m depth, 6.xi.2019, coll. JGH). **Curaçao**. RMNH.COEL.42431 (12°08′06″ N, 68°59′16″ W [point 63], Blue Wall, 10 m depth, 2.iv.2014, coll. BWH); MISE JDR170609-1-4 (12°07′17″ N, 68°58′09″ W [point 53], Carmabi, Hilton Hotel, 20 m depth, 9.vi.2017, coll. JDR); MISE JDR170610-4-30 (12°08′53″ N, 69°00′00″ W [point 65], Sint Michiel's Bay, 20 m depth, 10.vi.2017, coll. JDR); MISE JDR170612-7-45 (12°05′24″ N, 68°54′19″ W [point 50], Marie Pampoen, 31 m depth, 12.vi.2017, coll. JDR); MISE JDR170612-7-46 (12°05′24″ N, 68°54′19″ W [point 50], Marie Pampoen, 31 m depth, 12.vi.2017, coll. JDR); MISE JDR170614-11-67 (12°22′29″ N, 69°09′30″ W [point 76], Playa Kalki, 30 m depth, 14.vi.2017, coll. JDR); MISE NA (12°19′45″ N, 69°09′05″ W [point 74], Playa Jeremi, 12-38 m depth, 20.vi.2017, coll. JDR); ZMA.POR.14209 (12°07′23″ N, 68°58′14″ W [point 56], west of Piscadera Bay, 30 m depth, 11.v.1998, coll. M.J. de Kluijver); ZMA.POR.14245 (12°08′21″ N, 68°59′53″ W [point 64], Snake Bay, 20 m depth, 18.v.1998, coll. M.J. de Kluijver); ZMA.POR.16222 (coordinates and depth unknown, 12.i.1999, coll. H. Ranner); ZMA.POR.16312 (coordinates and depth unknown, vii.1992, coll. P. Willemsen); ZMA.POR.18396 (coordinates unknown, 27 m depth, 13.i.2003, coll. F.J. Parra-Velandia); ZMA.POR.22404 (12°07′32″ N, 68°58′27″ W [point 59], Buoy 1, north of Piscadera Bay, 15 m depth, 9.ii.1992, coll. R.W.M. Van Soest); ZMA.POR.3593 (12°04′29″ N, 68°52′50″ W [point 44], Jan Thiel Bay, 23–32 m depth, 16.xi.1975, coll. unknown); ZMA.POR.3623 (coordinates unknown, 10 m depth, xi.1975, coll. E. Westinga); ZMA.POR.3670 (12°07′45″ N, 68°58′51″ W [point 61], 500 m west of Piscadera Reef, 50 m depth, 7.xi.1975, coll. S. Weinberg); ZMA.POR.5703 (12°08′01″ N, 68°59′07″ W [point 62], Blue Bay, 55 m depth, 20.x.1984, coll. W.F. Hoppe & M.J.M. Reichert). **Sint Eustatius**. MISE JDR150610-2, JDR150610-3 (17°27′44.2″ N, 62°58′46.7″ W [point 97], Sta. EUX007, depth: 20 m depth, 10.vi.2015, coll. JDR); MISE JDR150610-17 (17°27′53.9″ N, 62°59′00.7″ W [point 101], Sta. EUX008, 16 m depth, 10.vi.2015, coll. JDR); MISE JDR150611-27 (17°27′42.1″ N, 62°58′41.2″ W [point 96], Sta. EUX009, 37 m depth, 11.vi.2015, coll. JDR); MISE JDR150611-64 (17°28′19.2″ N, 62°59′15.6″ W [point 107], Sta. EUX010, 13 m depth, 11.vi.2015, coll. JDR); MISE JDR150612-80 (17°30′57.4″ N, 62°59′21.6″ W [point 120], Sta. EUX011, 18 m depth, 12.vi.2015, coll. JDR); MISE JDR150612-97 (17°30′22.6″ N, 63°00′22.0″ W [point 117], Sta. EUX012, 13 m depth, 12.vi.2015, coll. JDR); MISE JDR150614-124 (17°27′50.9″ N, 62°59′06.8″ W [point 100], Sta. EUX015, 15 m depth, 14.vi.2015, coll. JDR); MISE JDR150615-135 (17°28′05.6″ N, 62°59′30.3″ W [point 104], Sta. EUX016, 21 m depth, 15.vi.2015, coll. JDR); MISE JDR150615-137 (17°28′05.6″ N, 62°59′30.3″ W [point 104], Sta. EUX016, 20 m depth, 15.vi.2015, coll. JDR); MISE JDR150616-145 (17°28′13.6″ N, 62°59′30.2″ W [point 106], Sta. EUX019, 19 m depth, 16.vi.2015, coll. JDR).

Photographic records (*n* = 28). In situ: Specimens RMNH.COEL.42431, MISE JDR150610-2, JDR150610-3, JDR150610-17, JDR150611-27, JDR150611-64, JDR150612-80, JDR150612-97, JDR150614-124, JDR150615-135, JDR150615-137, JDR150616-145, JDR170609-1-4, JDR170610-4-30, MISE JDR170612-7-45, JDR170612-7-46, JDR170614-11-67, JGH191106-2-1; Preserved: Specimens ZMA.POR.3593, 3623, 3670, 5703, 14209, 14245, 16222, 16312, 18396, 22404.

Description as in West [26]: Colonies with polyps regularly distributed over the sponge surface, but clusters formed by two or more polyps are also present; the distribution of polyps is variable depending on the host sponge. The polyp dimensions in living specimens are 1 mm in high and 1.5 mm in diameter; completely retracted polyps are mammiform, rising little above the surface of the coenenchyme, the coenenchyme surrounding the polyps is rarely more than 2 mm. Capitular ridges are 12 in number, mesenteries are 24 in number; tentacles arranged in two cycles and 24 in number with up to 1 mm in length in living and expanded specimens. The polyps present abundant pigmentation and a dense concentration of sponge spicules and calcareous sand grains, with an overall dark maroon color.

Recent and other previous records: Barbados [66], Colombia [67], Curaçao [24], Dominica [66], Puerto Rico [26], Tobago, USA (Navassa) [66], and Venezuela [70].

Remarks: As with other *Bergia* congeners listed above, this species appears in most literature as *P. puertoricense* as the genus *Bergia* was only recently resurrected [71]. *B. puertoricense* was found associated to *Petrosia* (*Petrosia*) *weinbergi* van Soest, 1980, *Petrosia* (*Petrosia*) aff. *weinbergi*, *Agelas clathrodes* (Schmidt, 1870), *Agelas conifera* (Schmidt, 1870), *Agelas* cf. *conifera* (Schmidt, 1870), *Svenzea zeai* (Alvarez, van Soest & Rützler, 1998), *Topsentia* sp., and *Xestospongia* sp. in Curaçao; and with *Svenzea zeai* in Sint Eustatius.

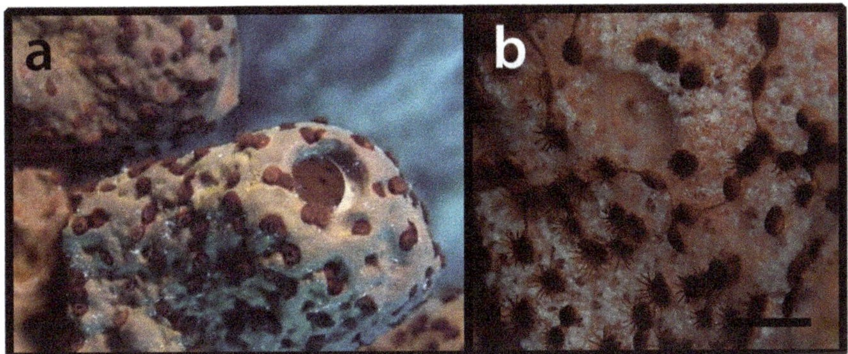

Figure 6. *Bergia puertoricense* in situ; (**a**) specimen MISE JDR170609-1-4 from Carmabi, Hilton Hotel [point 53], Curaçao, depth = 20 m, and (**b**) image (not collected) from Red Slave, Bonaire, depth = 30 m. Scale bar in (**b**) = approximately 1 cm.

Genus *Parazoanthus* Haddon & Shackleton, 1891 [50]

3.2.6. *Parazoanthus swiftii* (Duchassaing de Fonbressin & Michelotti, 1860) [23] (Figure 7)

Specimens examined (*n* = 22). **Bonaire.** MISE JDR191024-2-1 (12°7′53.28″ N, 68°16′59.76″ W [point 32], Corporal Meiss, 20 m depth, 24.x.2019, coll. JDR); MISE JDR191029-1-1 (12°13′24.42″ N, 68°24′13.38″ W [point 22], Taylor Made, 37 m depth, 29.x.2019, coll. JDR); MISE JDR191101-2-3 (12°15′49.8″ N, 68°24′49.2″ W [point 25], Boka Slaagbaai, 8 m depth, 1.xi.2019, coll. JDR). **Curaçao.** RMNH.COEL.42432 (12°06′33″ N 68°57′15″ W [point 52], Santa Marta, Water Factory, 20 m depth, 27.iii.2014, coll. BWH); MISE JDR170609-2-6 (12°06′33″ N, 68°57′15″ W [point 52], Santa Marta, Water Factory, 21 m depth, 9.vi. 2017, coll. JDR); MISE JDR170610-4-32 (12°08′53″ N, 69°00′00″ W [point 65], Sint Michiel's Bay, 19 m depth, 10.vi.2017, coll. JDR); MISE JDR170614-11-66 (12°22′29″ N, 69°09′30″ W [point 76], Playa Kalki south, 40 m depth, 14.vi.2017, coll. JDR); ZMA.POR.5839 (12°07′38″ N, 68°58′39″ W [point 60], Buoy 2, north of Piscadera Bay, 10 m depth, 2.i.1981, coll. R.W.M. Van Soest); ZMA.POR.10110 (coordinates and depth unknown, 1992, coll. P. Willemsen). **Saba.** ZMA.POR.15667 (17°36′29.7″ N, 63°15′07.6″ W [point 79], 800 m off Fort Bay, depth unknown, 12.iii.1986, coll. J. Vermeulen); RMNH.COEL.17763 (17°37′05″ N, 63°15′26″ W [point 81], Sta. LUY-021, between Fort Bay and Ladder Point, depth unknown, 10.iii.1986, coll. JCH). **Sint Eustatius.** MISE JDR150612-85, JDR150612-86 (17°30′57.4″ N, 62°59′21.6″ W [point 120], Sta. EUX011, 14–15 m depth, 12.vi.2015, coll. JDR); MISE JDR150612-94, JDR150612-98, JDR150612-101 (17°30′22.6″ N, 63°00′22.0″ W [point 117], Sta. EUX012, 14–15 m depth, 12.vi.2015, coll. JDR); MISE JDR150614-116, JDR150614-118 (17°29′00.6″ N 62°59′52.9″ W [point 96], Sta. EUX014, 21–22 m depth, 14.vi.2015, coll. JDR); MISE JDR150614-120 (17°29′00.6″ N, 62°59′52.9″ W [point 115], Sta. EUX014, 16 m depth, 14.vi.2015, coll. JDR); MISE JDR150617-150 (17°28′48.3″ N, 62°59′39.4″ W [point 112], Sta. EUX020, 17 m depth, 17.vi.2015, coll. JDR); MISE JDR150618-159 (17°27′56.6″ N, 63°00′07.2″ W [point 102], Sta. EUX022, 27 m depth, 18.vi.2015, coll. JDR); MISE JDR150619-165 (17°31′35.7″ N, 62°59′35.3″ W [point 121], Sta. EUX024, 29 m depth, 19.vi.2015, coll. JDR).

Photographic records (*n* = 20). In situ: Specimens RMNH.COEL.42432, MISE JDR150612-85, JDR150612-86, JDR150612-94, JDR150612-98, JDR150614-116, JDR150614-118, JDR150614-120, JDR150617-150, JDR150618-159, JDR150619-165, JDR170609-2-6, JDR170610-4-32, JDR170614-11-66, JDR191024-2-1, JDR191029-1-1, JDR191101-2-3. Preserved: Specimens ZMA.POR.5839, 15667, 10110.

Description as in Duchassaing de Fonbressin & Michelotti [23]: Small species with brownish/orange coloration, growing on the surface of the sponge in a linear pattern. The lines are formed by polyps connected to each other by short propagules. The lines are generally composed of 2 to 7 polyps, however solitary polyps and clusters of three to four polyps with no particular linear distribution are also present. Polyps are 1 mm in height and diameter and are not immersed in the sponge tissues.

Recent and other previous records: Ascension Islands [9,73], Barbados [65,66], Brazil [1,66], Colombia [67], Cuba [74], Curaçao [24,66], Dominica [66,75], Jamaica [68,76,77], Panama [66,78], Puerto Rico [26], Saint Thomas [23], Tobago, US Virgin Islands, USA (Georgia, Florida) [66], and Venezuela [70].

Remarks: This species has been shown to be closely related to *P. darwini* from the Galapagos in the Eastern Pacific via molecular studies [61]. *P. swiftii* was found associated with *Topsentia ophiraphidites* (Laubenfels, 1934) and *Topsentia* sp. in Curaçao [24]; with *Dragmacidon reticulatum* (Ridley & Dendy, 1886) in Saba; and with *Iotrochota birotulata* (Higgin, 1877) in Sint Eustatius.

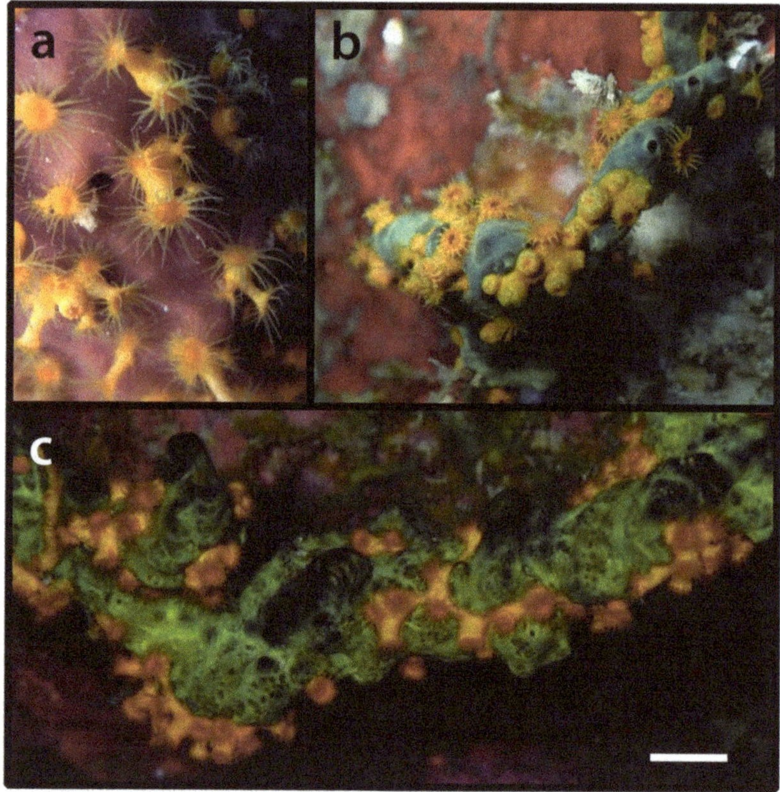

Figure 7. In situ images of *Parazoanthus swiftii* (**a**) RMNH.COEL.42432 from Santa Marta, Water Factory, Curaçao [point 52], depth = 20 m, (**b**) MISE JDR170610-4-32 from Sint Michiel's Bay [point 65], Curaçao, depth = 19 m, and (**c**) image (not collected) from Bari Reef, Bonaire, depth = 18 m. Scale bar in (**c**) = approximately 1 cm.

3.2.7. Parazoanthidae? sp. (Figure 8)

Specimens examined (*n* = 1). **Curaçao.** RMNH.COEL.40264 (12°07′45″ N, 68°58′51″ W [point 61], north of Piscadera Bay, depth unknown, 11.ix.1972, coll. JCH).

Photographic records (*n* = 1). Preserved: specimen RMNH.COEL.40264.

Remarks: The specimen RMNH.COEL.40264 was poorly preserved and was not possible to confidently identify to species level. While the polyp arrangement and size resembled preserved specimens of *P. swiftii*, there is also the possibility it is an *U. parasiticus* specimen. Future examination of the host sponge species should help more confidently identify this specimen as these zoantharian species do not have overlapping host species.

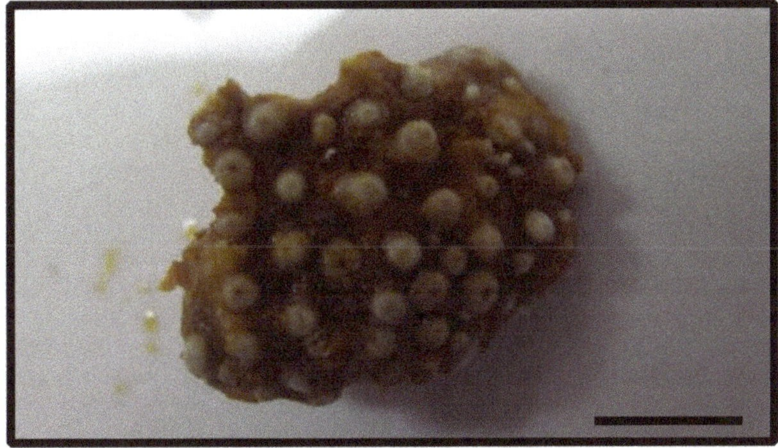

Figure 8. Preserved specimen RMNH.COEL.40264 Parazoanthidae? sp. from north of Piscadera Bay [point 61], Curaçao, depth = unknown. Scale bar = approximately 1 cm.

3.2.8. *Parazoanthus atlanticus* sp. n. (Figure 9)

http://zoobank.org/urn:lsid:zoobank.org:act:151E1AAA-7CAD-46AF-83B6-BFC0D0E0C931

Synonymy: *Parazoanthus* sp. 269 *sensu* Reimer et al. 2010 [4] (p. 162, Figure 2e)

Etymology: "atlanticus" in reference to the wide Atlantic distribution of this species, known from the Cape Verde Islands (East Atlantic) and Curaçao and Bonaire in the Caribbean.

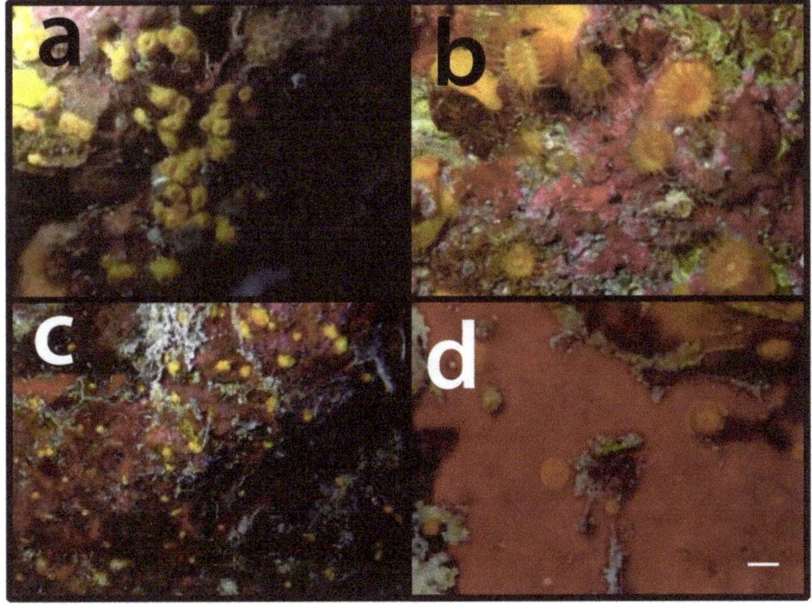

Figure 9. In situ images of *Parazoanthus atlanticus* sp. n. (**a**) specimen MISE JDR170613-10-61 from

Caracas Bay, Tugboat [point 42], Curaçao, depth = 18 m, (**b**) specimen MISE JDR170616-13-76 from north of Blue Bay [point 62], Curaçao, depth = 34 m, (**c**) specimen MISE JDR191103-1-1 from Small Wall [point 14], Bonaire, depth = 16 m, and (**d**) specimen MISE JDR191103-1-1 from Small Wall, Bonaire, depth = 16 m, with presence of much smaller *Umimayanthus* sp. Note the small sizes of the colonies formed by *P. atlanticus* sp. n. and the high frequency of solitary polyps. Scale bar in (**d**) = approximately 3 mm.

Material examined: Type locality Curaçao, Director's Bay [point 40], 12°03′59″ N, 68°51′38″ W (Table S1). **Holotype**: NSMT-Co 1706 (12°03′59″ N, 68°51′38″ W [point 40], Director's Bay, Curaçao, 27 m depth, 13.vi.2017, coll. JDR). **Paratype 1**: RMNH.COEL.42433 (12°08′06″ N, 68°59′16″ W [point 63], Blue Wall, Curaçao, on the ceiling of a cave at 10 m depth, 2.iv.2014, coll. BWH). **Paratype 2**: NSMT-Co 1707 (12°03′59″ N, 68°51′38″ W [point 40], Director's Bay, 20 m depth, 13.vi.2017, coll. JDR). **Other material** (*n* = 5). Other specimens are deposited in the Molecular Invertebrate Systematics and Ecology (MISE) Laboratory collection at the University of the Ryukyus, Nishihara, Okinawa, Japan. **Bonaire.** MISE JDR191029-1-3 (12°13′24.42″ N, 68°24′13.38″ W [point 22], Taylor Made, 25 m depth, 29.x.2019, coll. JDR); MISE JDR191103-1-1 (12°10′41.1″ N, 68°17′32.34″ W [point 14], Small Wall, 16 m depth, 3.xi.2019, coll. JDR). **Curaçao.** MISE JDR170613-10-60 (12°04′05″ N, 68°51′44″ W [point 42], Caracas Bay, Tugboat, 29 m depth, 13.vi.2017, coll. JDR); MISE JDR170613-10-61 (similar but at 18 m depth); MISE JDR170616-13-76 (12°08′01″ N, 68°59′07″ W [point 62], north of Blue Bay, 34 m depth, 16.vi.2017, coll. JDR).

Photographic records (*n* = 8). **Bonaire.** In situ: Specimens MISE JDR191029-1-3, JDR191103-1-1. **Curaçao.** In situ: Specimens RMNH.COEL.42433, NSMT-Co 1706, NSMT-Co 1707, MISE JDR170613-10-60, JDR170613-10-61, JDR170616-13-76.

Sequences: All sequences were deposited in GenBank with accession numbers MT102222-MT102229 for the mitochondrial cytochrome oxidase subunit I region (COI-mtDNA), MT103533-MT103540 for mitochondrial ribosomal subunit 16S ribosomal DNA, and MT103524-MT103531 for the nuclear ribosomal internal transcribed spacer sequences (ITS-rDNA).

Description. Size: Preserved polyps are on average 3.117 mm ± 0.640 mm (σ^2 = 0.427, *n* = 24 polyps) in diameter and 1.446 mm ± 0.569 mm (σ^2 = 0.337, *n* = 24 polyps) in height. All measurements were performed on closed polyps of specimens preserved in 99% ethanol.

Morphology: *Parazoanthus atlanticus* sp. n. presents a bright yellow color, almost orange, in all collected material and in situ. The polyps have approximately 24 to 30 tentacles. Colonies generally consist of clusters of three polyps scattered over the sponge surface, but single polyps and groups of up to 14 polyps were also observed. Polyps were solitary or connected to each other by the stolon over the sponge surface. Distance between polyps was variable and no noticeable pattern was found.

Cnidae: All cnidocyte categories previously reported in Zoantharia [32] were found, however holotrichs and p-mastigophores were particularly low in frequency in all examined tissues (tentacles, column, pharynx and mesenterial filaments); p-mastigophores were only found in mesenterial filaments, and holotrichs medium were found only in the pharynx. Spirocysts were absent in column and mesenterial filaments. For details on sizes, lengths, and widths of each cnidocyte categories, refer to Table 3 and Figure 10.

Table 3. Results of the cnidocyte analyses of all categories found per examined tissue. Notice the differential distribution and frequency of each cnidocyte category across tissues.

Sample ID: NSMT-Co 1706		Length (Min-Max, Average) μm	Width (Min-Max, Average) μm	*n*
Tentacles	Spirocysts	13–29, 20	2–6, 3.5	222
	Holotrichs (L)	32	15	1
	Bastrichs and microbasic b-mastigophores	14–23, 19.8	2–5, 4.3	24
	Microbasic p-mastigophores	-	-	-
Column	Spirocysts	-	-	-
	Holotrichs (L)	20–47, 29.2	11–15, 13.1	16
	Bastrichs and microbasic b-mastigophores	-	-	-
	Microbasic p-mastigophores	-	-	-
Pharynx	Spirocysts	23	3	1
	Holotrichs (M)	17	8	1
	Bastrichs and microbasic b-mastigophores	14–18, 16.1	2–4, 3.3	17
	Microbasic p-mastigophores	-	-	-
Filaments	Spirocysts	-	-	-
	Holotrichs (L)	25–31, 28.4	9–16, 13.5	32
	Bastrichs and microbasic b-mastigophores	-	-	-
	Microbasic p-mastigophores	12–20, 16	3–6, 4.8	10

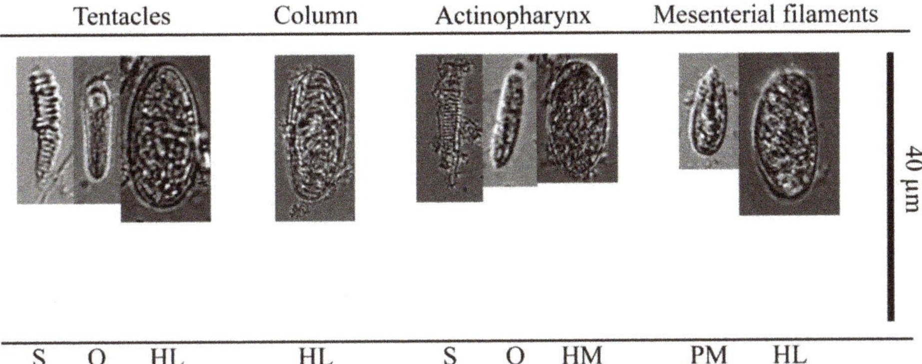

Figure 10. Images of all cnidae categories and average size of found across different tissues in polyps of *P. atlanticus* sp. n.; S = Spirocysts, O = basitrich & microbasic b-mastigophores, HL = Holotrichs (L), HM = Holotrichs (M), and PM = microbasic p-mastigophores.

Differential diagnosis: *Parazoanthus atlanticus* sp. n. can be distinguished from *P. anguicomus* (Norman, 1869) [79], *P. haddoni* Carlgren, 1913 [54], *P. antarcticus* Carlgren, 1927 [80], *P. aruensis* Pax, 1911 [81], *P. elongatus* McMurrich, 1904 [82], and *P. juan-fernandezii* Carlgren, 1922 [83] by polyp diameter and numbers of tentacles, which are larger in all the above species (see also [71]). As well, *P. darwini* Reimer & Fujii, 2010 [61] and *P. lividum* Cutress, 1971 [84] differ in distribution ranges, being only found in the South and East Pacific; additionally, multiple morphological characteristics set *P. lividum* apart, including polyp size and colonies formed by polyps organized in a band-like arrangement [84]. Although the descriptions of *P. axinellae* (Schmidt, 1862) [85] and *P. capensis* Duerden, 1907 [86] overlap with *P. atlanticus* sp. n. regarding the number of tentacles and polyp diameters, both *P. axinellae* and *P. capensis* associate with a different sponge species. *P. axinellae* was reported associated with *Axinella verrucosa*, *A. damicornis*, *Petrosia ficiformis*, and *Hippospongia communis*; while for *P. capensis*, the associated sponge species remains unknown, but the sponges are arboresent/branching in shape [86]. On the other hand, the sponges associated to *P. atlanticus* sp. n., although not identified, are all encrusting in morphology. *P. capensis* from South Africa was described with pale yellow polyps with colorless tentacles [86], different from the brighter yellowish coloration in *P. atlanticus* sp. n. Within genus *Parazoanthus*, the species that most resembles *P. atlanticus* sp. n. in colony shape and polyp color is *P. swiftii* (Duchassaing de Fonbressin & Michelotti, 1860) [23], however it may be slightly differentiated by polyp size; 2.5 mm in diameter for *P. swiftii* and 3.1 mm for *P. atlanticus* sp. n. Additionally, in colonies of *P. swiftii*, solitary polyps are only exceptionally found while in *P. atlanticus* sp. n., solitary polyps are relatively frequently observed. As well, while *P. swiftii* is often found in environments with abundant light exposure and is associated to a wide range of sponges, *P. atlanticus* sp. n. has only been found in cave-like environments and exclusively associates with encrusting sponges.

Encrusting sponges in caves and cracks of Bonaire have been reported to have high levels of diversity and contain many cryptic species [87]. Accurate identification of the host encrusting sponges of *P. atlanticus* sp. n. in the future would further help characterize differences with closely related species. As well, microanatomical analyses (e.g., [53]) should also help further differentiate *P. atlanticus* sp. n. from closely related species.

Phylogenetic analyses using the sequences of the ITS-rDNA region also support *P. atlanticus* sp. n. as a monophyly with complete support in both ML and BPP analyses. Similar tree topologies for the COI-mtDNA and 16S-rDNA analyses were seen but with weaker support (Figure S1). When the sequences of COI-mtDNA, 16S-rDNA, and ITS-rDNA were concatenated, the monophyly of *P. atlanticus* sp. n. was moderately supported (ML = 71%, BPP = 0.86; Figure 11). Although the phylogenetic position of *P. atlanticus* sp. n. within genus *Parazoanthus* remains uncertain, both the ITS-rDNA and concatenated phylogenies weakly support *P. atlanticus* sp. n. as a basal or sister clade to a clade formed by *P. axinellae*, *P. anguicomus*, and *P. capensis*. Remarkably, in the COI-mtDNA region, a single nucleotide substitution from "C" to "T" at position 178 of our alignment was found to be unique to *P. atlanticus* sp. n. across all species of sponge-associated zoantharian in genera *Bergia*, *Parazoanthus* and *Umimayanthus*; several substitutions and indels were also found to be unique to *P. atlanticus* sp. n. in ITS-rDNA and 16S-rDNA regions.

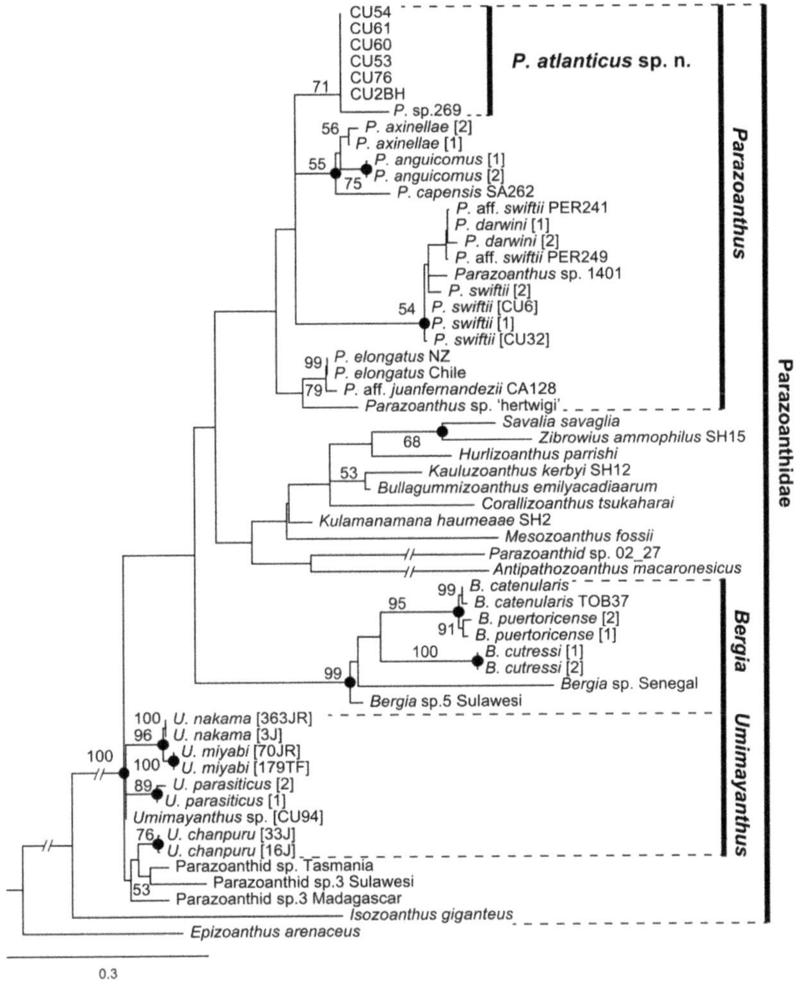

Figure 11. Maximum likelihood (ML) phylogenetic reconstruction using the concatenated alignment of one nuclear DNA region ITS-rDNA, and two mitochondrial regions 16S-rDNA and COI mt-DNA. Values on branches represent bootstrap support >50 from maximum likelihood analyses and black circles indicate Bayesian posterior probabilities >95%. Note the clear monophyly form by *P. atlanticus* sp. n. in reference to other species in genus *Parazoanthus*. CU53 = NSMT-Co 1706; CU54 = NSMT-Co 1707; CU60 = MISE JDR170613-10-60; CU61 = MISE JDR170613-10-61; CU76 = MISE JDR170616-13-76; CU2BH = RMNH.COEL.42433; CU6 = MISE JDR170609-2-6; CU32 = MISE JDR170610-4-32; CU94 = MISE JDR170619-20-94.

Distribution. Localities and islands recorded in this study: **Bonaire.** Taylor Made [point 22] and Small Wall [point 14]. **Curaçao.** Director's Bay [point 40], Caracas Bay [point 42], Piscadera Bay [point 61], Blue Bay [point 62], and Blue Wall [point 63].

Recent and other previous records: *P. atlanticus* sp. n. was previously reported in Reimer et al. [4] as *Parazoanthus* sp. 269; the specimen was collected at Danger, Tarrafal Bay, Santiago I. in Cape Verde from a submerged cave at 20 m depth.

Genus *Umimayanthus* Montenegro, Sinniger & Reimer, 2015 [88]

3.2.9. Umimayanthus parasiticus (Duchassaing de Fonbressin & Michelotti, 1860) [23] (Figure 12)

Specimens examined (*n* = 78). **Bonaire**. RMNH.COEL.40255 (12°05′40″ N, 68°17′06″ W [point 7], Inlet Salt Lake West Coast, 1–2 m depth, 10.iv.1978, coll. JCH). **Curaçao**. RMNH.COEL.42434 (12°06′33″ N, 68°57′15″ W [point 52], Santa Marta, Water Factory, 20 m depth, 27.iii.2014, coll. BWH); MISE JDR170609-1-1, JDR170609-1-2, JDR170609-1-5 (12°07′17″ N, 68°58′09″ W [point 53], Carmabi, Hilton Hotel, 15–32 m depth, June 9, 2017, coll. JDR); MISE JDR170613-9-55 (12°03′59″ N, 68°51′38″ W [point 40], Director's Bay, 20 m depth, 13.vi.2017, coll. JDR); MISE JDR170619-21-97 (12°13′55″ N, 69°05′44″ W [point 67], Playa Cas Abao, 20 m depth, 19.vi.2017, coll. JDR); MISE NA (12°19′45″ N, 69°09′05″ W [point 74], Playa Jeremi, 24–38 m depth, 20.vi.2017, coll. JDR); RMNH.COEL.40239 (12°07′45″ N, 68°58′51″ W [point 61], Buoy 2, north of Piscadera Bay, 5 m depth, 9.ix.1972, coll. JCH); RMNH.COEL.40258 (12°04′47″ N, 68°50′25″ W [point 47], Spaanse Water east side, depth unknown, 9.vi.1971, coll. JCH); RMNH.COEL.40269 (12°04′25″ N, 68°50′37″ W [point 43], Spaanse Water near Fosfaatberg, shallow, 12.xi.1972, coll. JCH); ZMA.POR.10756 (12°03′48″ N, 68°51′04″ W [point 38], Spaanse Water, depth unknown, 1989, coll. M. Kuenen); ZMA.POR.14194 (12°04′49″ N, 68°53′18″ W [point 48], Jan Thiel, Curaçao, 38 m depth, 10.v.1998, coll. M.J. de Kluijver); ZMA.POR.14198 (12°04′29″ N, 68°52′50″ W [point 44], Jan Thiel Bay, 24 m depth, 10.v.1998, coll. M.J. de Kluijver); ZMA.POR.14271 (12°07′23″ N, 68°58′14″ W [point 56], west of Piscadera Bay, 34 m depth, 22.v.1998, coll. M.J. de Kluijver); ZMA.POR.14353 (12°07′17″ N, 68°58′09″ W [point 53], Carmabi, Hilton Hotel, 18 m, 21.v.1998, coll. R. Gomez); ZMA.POR.20902 (12°6′16.6″ N, 68°56′36.6″ W [point 77], Superior Producer wreck, 33 m depth, 23.v.1998, coll. R. Gomez); ZMA.POR.20913 (12°6′26.6″ N, 68°56′51.7″ W [point 78], Holiday Beach, 19 m depth, 30.i.2000, coll. M.J. de Kluijver); ZMA.POR.20914 (12°6′26.6″ N, 68°56′51.7″ W [point 78], Holiday Beach, depth unknown, 18.ii.2000, coll. R. Gomez); ZMA.POR.22251 (12°07′20″ N, 68°58′08″ W [point 54], Carmabi, house reef, 22 m depth, 1.v.1991, coll. P. Willemsen); ZMA.POR 3304, 3634 (12°04′03″ N, 68°51′10″ W [point 41], Santa Barbara Beach, Spanish Lagoon, 3 m depth, 17.i.1974, 18.i.1974, coll. J.H. Stock); ZMA.POR.3305, 3306 (12°07′45″ N, 68°58′51″ W [point 61], north of Piscadera Reef, 32–40 m depth, 19.xii.1973, 22.iii.1974, coll. J.H. Stock); ZMA.POR.3315 (12°07′45″ N, 68°58′51″ W [point 61], north of Piscadera Reef, 10-18 m depth, 21.xi.1973, coll. J.H. Stock); ZMA.POR.3486 (12°08′01″ N, 68°59′07″ W [point 62], Blue Bay, 3 m depth, 17.x.1958, coll. J.H. Stock); ZMA.POR.3581 (12°07′20″ N 68°58′08″ W [point 54], Carmabi, House Reef, 10–25 m depth, 10.xii.1975, coll. E. Westinga); ZMA.POR.3600, 3647 (12°08′01″ N, 68°59′07″ W [point 62], Blue Bay, 20–30 m depth, xi.1975, coll. S. Weinberg & E. Westinga); ZMA.POR.3601 (12°07′23″ N, 68°58′14″ W [point 56], west of Piscadera Bay, 15 m depth, 13.xi.1975, coll. unknown); ZMA.POR.3609, 3648, 3653, 3644 (12°07′20″ N, 68°58′08″ W [point 54], Carmabi, House Reef, 11–20 m depth, 14.xi.1978, 16.xi.1978, coll. unknown); ZMA.POR.3614 (12°08′01″ N, 68°59′07″ W [point 62], Blue Bay, 15–20 m depth, xi.1975, coll. S. Weinberg & E. Westinga); ZMA.POR.3646 (12°07′32″ N, 68°58′27″ W [point 59], Buoy 1, north of Piscadera Bay, 40 m depth, 22.iii.1974, coll. J.H. Stock); ZMA.POR.3877 (12°07′45″ N, 68°58′51″ W [point 61], north of Piscadera Reef, 35 m depth, 13.xi.1975, coll. unknown). **Saba**. ZMA.POR 15745 (17°36′29.7″ N, 63°15′07.6″ W [point 79], 800 m off Fort Bay, Saba, depth unknown, 12.iii.1986, coll. J. Vermeulen). **Saba Bank**. RMNH.POR.5102, 5152, 5153 (17°33′00″ N, 63°22′00″ W [point 94], Sta. LUY-124, northeast side, 24 m depth, 12.xi.1972, coll. Luymes Exp.); RMNH.POR.5103 (17°12′00″ N, 63°40′00″ W [point 85], Sta. LUY-111, southwest side, 28 m depth, 24.v.1972, coll. Luymes Exp.); RMNH.POR.5104 (17°14′00″ N, 63°34′00″ W [point 86], Sta. LUY-055, northeast side, 39 m depth, 15.x.1972, coll. Luymes Exp.); RMNH.POR.5105 (17°29′00″ N, 63°13′00″ W [point 92], Sta. LUY-144, east side, 16 m depth, 14.xi.1972, coll. Luymes Exp.); RMNH.POR.5148 (17°23′00″ N, 63°45′00″ W [point 90], Sta. LUY-069, 44 m depth, 17.x.1972, coll. Luymes Exp.); RMNH.POR.5149 (17°33′00″ N, 63°27′00″ W [point 95], Sta. LUY-126, 36 m depth, 12.xii.1972, coll. Luymes Exp.); RMNH.POR Sta-146 (17°26′00″ N, 63°12′00″ W [point 91], Sta. LUY-146, east-slope, 30 m depth, 14.vi.1972, coll. Luymes Exp.). **Sint Eustatius**. MISE JDR150610-1, JDR150610-5, JDR150610-8 (17°27′44.2″ N, 62°58′46.7″ W [point 97], Sta. EUX007, 19–21 m depth, 10.vi.2015, coll. JDR); MISE JDR150610-10, JDR150610-11, JDR150610-13, JDR150610-14, JDR150610-22 (17°27′53.9″ N, 62°59′00.7″ W [point 101], Sta. EUX008, Sint Eustatius, 16–17 m depth, 10.vi.2015, coll. JDR); MISE JDR150610-25, JDR150610-26 (17°27′53.9″ N, 62°59′00.7″ W [point 101], Sta. EUX008, depth unknown, 10.vi.2015, coll. JDR); MISE JDR150611-28 (17°27′42.1″ N, 62°58′41.2″ W [point 96], Sta. EUX009, 34 m depth, 11.vi.2015, coll. JDR); MISE JDR150611-32, JDR150611-41, JDR150611-63, JDR150611-76 (17°28′19.2″ N, 62°59′15.6″ W [point 107], Sta. EUX010, 11–13 m depth, 11.vi.2015, coll. JDR); MISE JDR150612-78, JDR150612-82, JDR150612-83 (17°30′57.4″ N, 62°59′21.6″ W [point 120], Sta. EUX011, 10–17 m depth, 12.vi.2015, coll. JDR); MISE JDR150612-91, JDR150612-99 (17°30′22.6″ N, 63°00′22.0″ W [point 117], Sta. EUX012, 12–15 m depth, 12.vi.2015, coll. JDR); MISE JDR150613-109 (17°28′56.3″ N, 62°59′20.3″ W [point 113], Sta. EUX013, 3 m depth, 13.vi.2015, coll. JDR); MISE JDR150614-117 (17°29′00.6″ N, 62°59′52.9″ W [point 115], Sta. EUX014, 22 m depth, 14.vi.2015, coll. JDR); MISE JDR150614-126, JDR150614-133 (17°27′50.9″ N, 62°59′06.8″ W [point 100], Sta. EUX015, 15–17 m depth, 14.vi.2015, coll. JDR); MISE JDR150615-134 (17°28′05.6″ N, 62°59′30.3″ W [point 104], Sta. EUX016, 21 m depth, 15.vi.2015, coll. JDR); MISE JDR150615-141 (17°28′31.5″ N, 62°59′33.6″ W [point 108], Sta. EUX017, 19 m depth, 15.vi.2015, coll. JDR); MISE JDR150616-143, JDR150616-144 (17°28′13.6″ N, 62°59′30.2″ W [point 106], Sta. EUX019, 19 m depth, 16.vi.2015, coll. JDR); MISE JDR150619-162 (17°31′35.7″ N, 62°59′35.3″ W [point 121], Sta. EUX024, 28 m depth, 19.vi.2015, coll. JDR); RMNH.POR.5099 (17°28′02.7″ N, 62°58′44.7″ W [point 103], Sta. LUY-121, <15 m depth, 9–10.xi.1972, coll. Luymes Exp.). **Sint Maarten**. RMNH.POR.5150 (18°04′00″ N, 63°06′00″ W [point 124], Sta. LUY-122, Baie Marigot, <15 m depth, 11.xi.1972, coll. Luymes Exp.).

Photographic records (*n* = 73). In situ: Specimens RMNH.COEL.42434, MISE JDR150610-1, JDR150610-5, JDR150610-8, JDR150610-10, JDR150610-11, JDR150610-13, JDR150610-14, JDR150610-22, JDR150611-28, JDR150611-32, JDR150611-41, JDR150611-76, JDR150612-78, JDR150612-82, JDR150612-83, JDR150612-91, JDR150612-99, JDR150613-109, JDR150614-117, JDR150614-126, JDR150615-134, JDR150615-141, JDR150616-143, JDR150616-144, JDR150619-162, JDR170609-1-1, JDR170609-1-2, JDR170609-1-5, JDR170613-9-55, JDR170619-21-97; Preserved: Specimens RMNH.COEL.40239, 40255, 40258, 40269, RMNH.POR.5099, 5102, 5103, 5104, 5105, 5148, 5149, 5150, 5152, 5153, Sta-146, ZMA.POR.3304, 3305, 3306, 3315, 3486, 3581, 3600, 3601, 3609, 3614, 3634, 3644, 3646, 3647, 3648, 3653, 3877, 10756, 14194, 14198, 14271, 14353, 15745, 20902, 20913, 20914, 22251.

Description as in Duchassaing de Fonbressin & Michelotti [23] and West [26]: A very small species living in sponges, and given its parasitic habits is similar to genus *Bergia* (note this species is now not thought to be generally parasitic [89]). It is a zoantharian without tissues reinforced with sand deposits. The propagules are basilar. The oral disk, including tentacles, is about 3/4 of the total polyp diameter. The length and diameter off expanded polyps are between 1 and 1.5 mm, the retracted polyps are mammiform in shape and rising little from above the sponge surface; the diameter of the coenenchyme surrounding the polyps is about 2 mm. Polyps are usually solitary or occasionally in groups of 2-3; the polyps present 14 ridges and 28 tentacles. The tentacles are yellow–brown in color, due to the symbiotic zooxanthellae.

Recent and other previous records: Bahamas [64], Barbados [65,66], Bermudas [90–93], Colombia [67], Cuba [74], Curaçao [24,66], Dominica [66,75], Mexico [94], Panama [66,78], Puerto Rico [26], Sint Eustatius [22], Saint Thomas [23], Tobago, US Virgin Islands [66], USA (Florida; Navassa) [64,66,95,96], Venezuela [70].

Remarks: Endemic to the Caribbean Sea, this species was the second most commonly observed zoantharian in the Dutch Caribbean, recorded to all islands with exception of Aruba. This species was placed in *Parazoanthus* until recent phylogenetic analyses followed by description of the genus *Umimayanthus* changed its placement [88]. *U. parasiticus* was found associated with *Callyspongia* (*Cladochalina*) *vaginalis* (Lamarck, 1814), *Niphates erecta* Duchassaing de Fonbressin & Michelotti, 1864, *N. amorpha* Van Soest, 1980, and *Svenzea zeai* in Curaçao ([24], this study); with *Callyspongia* (*Cladochalina*) *armigera* (Duchassaing de Fonbressin & Michelotti, 1864) in Saba; with *Niphates digitalis* (Lamarck,

1814), *Niphates erecta*, *Callyspongia* (*Cladochalina*) *vaginalis* in Saba Bank; with *Niphates digitalis* in Sint Eustatius; and with *Niphates amorpha* in Sint Maarten.

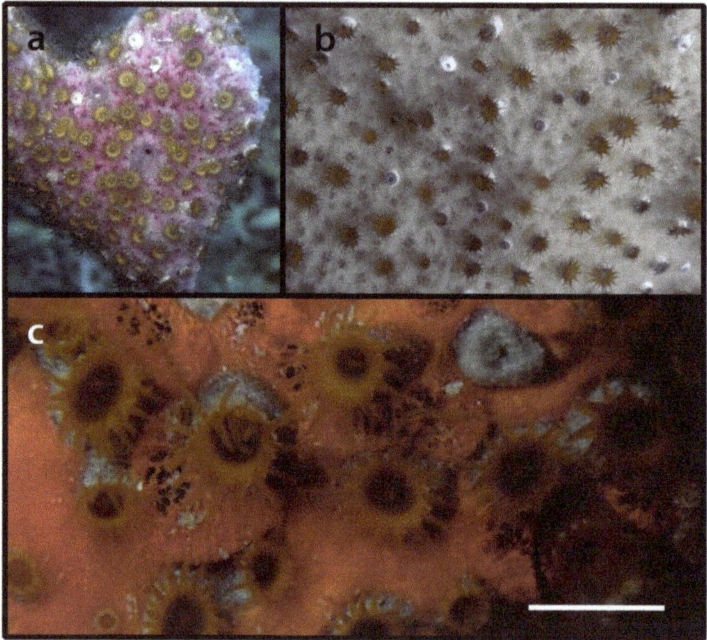

Figure 12. In situ images of *Umimayanthus parasiticus* (**a**) Specimen RMNH.COEL.42434 from Santa Marta, Water Factory [point 52], Curaçao, depth = 20 m, (**b**) image M0057481 (not collected) from Sta. EUX31 (Mushroom), Sint Eustatius, depth = 13 m, and (**c**) image PB045491 (not collected) from Front Porch, Bonaire, depth = 9 m. Scale bar in (**c**) = approximately 5 mm.

3.2.10. *Umimayanthus* sp. (Figure 13)

Specimens examined (*n* = 2). **Bonaire.** MISE JDR191026-1-1 (12°13′10.26″ N, 68°21′7.38″ W [point 20], Karpata, 37 m depth, 26.x.2019, coll. JDR). **Curaçao.** MISE JDR170619-20-94 (12°04′05″ N, 68°51′44″ W [point 42], Caracas Bay, Tugboat, 37 m depth, 19.vi.2017, coll. JDR).

Photographic records (*n* = 2). In situ: Specimens MISE JDR170619-20-94, MISE JDR191026-1-1.

Remarks: The specimens generally resemble *U. chanpuru* Montenegro, Sinniger & Reimer, 2015 [88] widely distributed across the Indo-Pacific region, but no similar species has been previously reported in the Atlantic Ocean or Caribbean Sea. Preliminary molecular analyses of the ITS-rDNA nuclear region indicate that specimen MISE JDR170619-20-94 is an undescribed species of genus *Umimayanthus* (GenBank Accession Number MT102227). Additional specimens and observations are needed to formally describe this species.

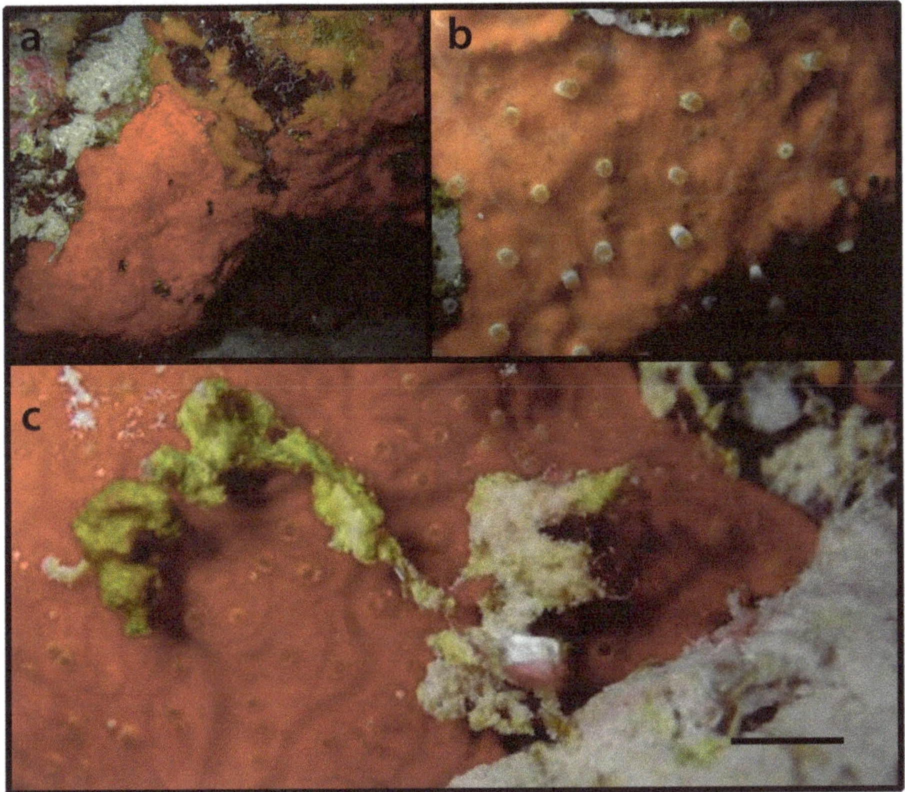

Figure 13. In situ images of *Umimayanthus* sp. (**a**) specimen MISE JDR170619-20-94 from Caracas Bay, Tugboat [point 42], Curaçao. Depth = 37 m, (**b**) close-up of same specimen, and (**c**) specimen MISE JDR191026-1-1 from Karpata, Bonaire, depth = 37 m. Scale bar in (**c**) = approximately 5 mm.

Family Epizoanthidae Delage & Hérouard, 1901 [51]
Genus *Epizoanthus* Gray, 1867 [72]

3.2.11. *Epizoanthus* sp. (Figure 14)

Specimens examined (*n* = 1). **Saba.** RMNH.COEL.40667 (17°49′00″ N, 63°16′00″ W [point 84], Sta. LUY-155, 980 m depth, 16.vi.1972, coll. Luymes Exp.).

Photographic records (*n* = 1). Preserved: Specimen RMNH.COEL.40667.

Remarks: Specimen RMNH.COEL.40667 does not resemble any known species in genus *Epizoanthus*, thus we consider it an undescribed species from the deep sea. Additional specimens and observations are needed to formally describe this species. Future collections of such material would do well to additionally collect detailed in situ images and information on host species (if present).

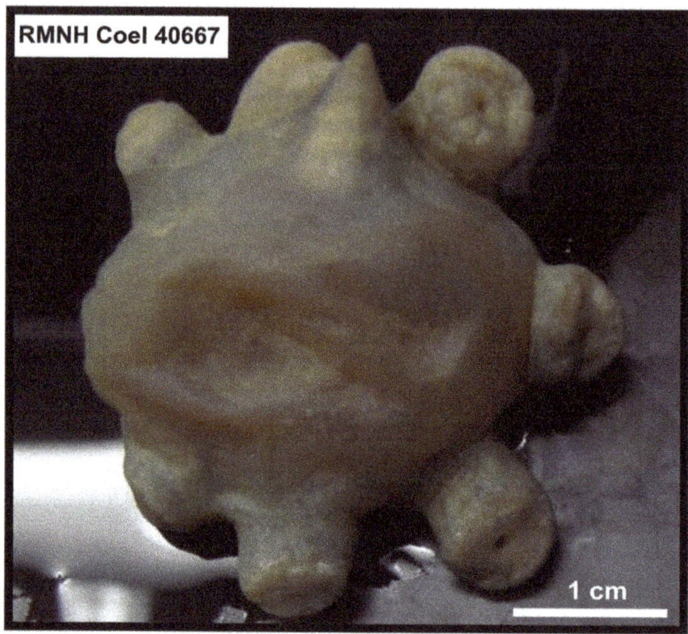

Figure 14. *Epizoanthus* sp. specimen RMNH.COEL.40667 from Sta. LUY-155 [point 84], Curaçao, depth = 980 m.

Family Hydrozoanthidae Sinniger, Reimer & Pawlowski, 2010 [58]
Genus *Hydrozoanthus* Sinniger, Reimer & Pawlowski, 2010 [58]

3.2.12. *Hydrozoanthus antumbrosus* (Swain, 2009) [62] (Figure 15)

Specimens examined (n = 14). **Bonaire.** MISE JDR191030-2-2 (12°4′52.92″ N, 68°13′55.56″ W [point 29], Baby Beach (Pretty Rough), 25 m depth, 30.x.2019, coll. JDR); MISE ML191108-1-1 (12°9′53.16″ N, 68°19′22.74″ W [point 34], Carl's Hill, Klein Bonaire, 20 m depth, 8.xi.2019, coll. Marianne Ligthart). **Curaçao.** MISE JDR170609-2-8 (12°06′33″ N, 68°57′15″ W [point 52], Santa Marta, Water Factory, 30 m depth, 9.vi.2017, coll. JDR). **Sint Eustatius.** MISE JDR150611-53, MISE JDR150611-61, MISE JDR150611-62 (17°28′19.2″ N, 62°59′15.6″ W [point 107], Sta. EUX010, 11 m depth, 11.vi.2015, coll. JDR); MISE JDR150612-92, JDR150612-96 (17°30′22.6″ N, 63°00′22.0″ W [point 117], Sta. EUX012, 15 m depth, 12.vi.2015, coll. JDR); MISE JDR150614-128 (17°27′50.9″ N, 62°59′06.8″ W [point 100], Sta. EUX015, 15 m depth, 14.vi.2015, coll. JDR); MISE JDR150615-142 (17°28′31.5″ N, 62°59′33.6″ W [point 108], Sta. EUX017, 19 m depth, 15.vi.2015, coll. JDR); MISE JDR150617-151 (17°27′50.4″ N, 62°59′15.0″ W [point 99], Sta. EUX021, 15 m depth, 17.vi.2015, coll. JDR); MISE JDR150617-152, JDR150617-154, JDR150617-155 (17°27′50.4″ N, 62°59′15.0″ W [point 99], Sta. EUX021, 15–17 m depth, 17.vi.2015, coll. JDR).

Photographic records (n = 13). In situ: Specimens MISE JDR170609-2-8, JDR150611-53, JDR150611-62, JDR150612-92, JDR150612-96, JDR150614-128, JDR150615-142, JDR150617-151, JDR150617-152, JDR150617-154, JDR150617-155, JDR191030-2-2, ML191108-1-1.

Description as in Swain [62]: The expanded polyps are dichromatic; coenenchyme, column, and oral disk are seal-brown and tentacles are golden. The polyps' column dimensions are 4.1 to 8.9 mm in length and 2.2 to 4.3 mm in diameter, and the oral disk is 2.7 to 4.8 mm in diameter. The contracted polyps are monochromatic, mammiform, 2.2 to 4.2 mm in diameter, and extending 3.3 to 9.9 mm above the surrounding coenenchyme. The capitulum has 15-19 distinct ridges. The tentacles are 30 to 38 in number, 1.9 to 5.0 mm long, and 0.4 to 0.7 mm diameter at the insertion point in the oral disk.

The colony present a thin and encrusting coenenchyme densely infiltrated with calcareous sediment and siliceous spicules. Polyps are separated by intervals of approximately 1.5 to 2.5 times a polyp diameter, often organized in an orthogonal or distichous arrangement with oral disk nearly parallel to the plane of the pinnate hydroid branches. The colony completely envelopes the central and secondary axial branches of the hydroid host, *Dentitheca dendritica* (Nutting, 1900), but usually not covering the pinnate branches where the hydroid zooids are located.

Recent and other previous records: Colombia, Dominica [62], Curaçao [24,62], Honduras [37,62], Panama, and Surinam [62].

Remarks: This species was locally common at sites around Bonaire, Curaçao, and Sint Eustatius where there was consistently strong current, which may be a prerequisite for the hydroid host species *Dentitheca* Stechow, 1920.

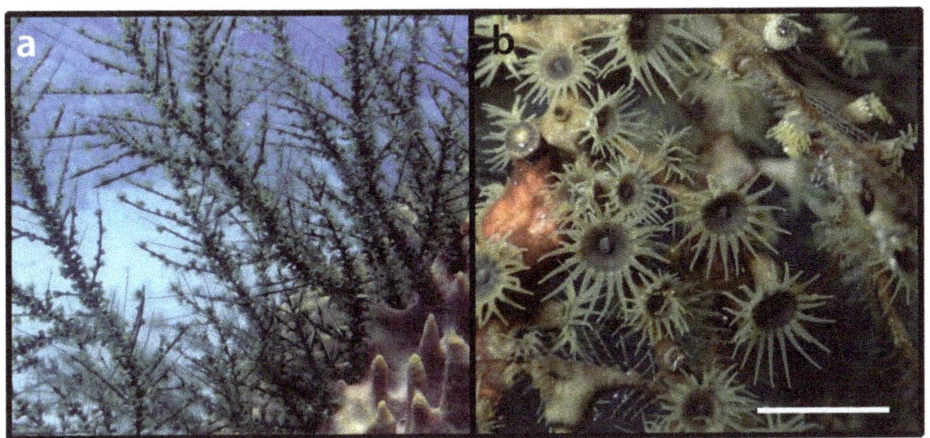

Figure 15. In situ images of *Hydrozoanthus antumbrosus*; (**a**) specimen MISE JDR150611-62 from Sta. EUX010 [point 107], Sint Eustatius, depth = 11 m, and (**b**) specimen MISE JDR170609-2-8 from Santa Marta, Water Factory [point 52], Curaçao, depth = 30 m. Scale bar in (**b**) = approximately 1 cm.

3.2.13. *Hydrozoanthus tunicans* (Duerden, 1900) [68] (Figure 16)

Specimens examined (*n* = 11). **Bonaire.** MISE JDR191030-2-1 (12°4′52.92″ N, 68°13′55.56″ W [point 29], Baby Beach (Pretty Rough), 25 m depth, 30.x.2019, coll. JDR). **Curaçao.** MISE JDR170609-2-7 (12°06′33″ N, 68°57′15″ W [point 52], Santa Marta, Water Factory, 30 m depth, 9.vi.2017, coll. JDR); RMNH.COEL.13201 (coordinates unknown, south Coast, 2–4 m depth, 9.iv.1972, coll. JCH). **Sint Eustatius.** MISE JDR150610-9 (17°27′53.9″ N, 62°59′00.7″ W [point 101], Sta. EUX008, 18 m depth, 10.vi.2015, coll. JDR); MISE JDR150612-93 (17°30′22.6″ N, 63°00′22.0″ W [point 117], Sta. EUX012, 14 m depth, 12.vi.2015, coll. JDR); MISE JDR150614-123 (17°27′50.9″ N, 62°59′06.8″ W [point 100], Sta. EUX015, 15 m depth, 14.vi.2015, coll. JDR); MISE JDR150615-138, JDR150615-139, JDR150615-140 (17°28′31.5″ N, 62°59′33.6″ W [point 108], Sta. EUX017, 19 m depth, 15.vi.2015, coll. JDR); MISE JDR150616-146 (17°28′13.6″ N, 62°59′30.2″ W [point 106], Sta. EUX019, 16 m depth, 16.vi.2015, coll. JDR); MISE JDR150617-153 (17°27′50.4″ N, 62°59′15.0″ W [point 99], Sta. EUX021, 17 m depth, 17.vi.2015, coll. JDR).

Photographic records (*n* = 10). In situ: Specimens MISE JDR150610-9, JDR150612-93, JDR150614-123, JDR150615-138, JDR150615-139, JDR150615-140, JDR150616-146, JDR150617-153, JDR170609-2-7, JDR191030-2-1.

Description as in Duerden [68]: Each colony consists of a thin coenenchyme with multiple polyps arising close to each other, covering the main stems and smaller branches of hydroids. On the smaller branches, the polyps are arranged in a distichous manner, in a plane at right angles to the pinnulae of the hydroid, and polyps at the two sides are either opposite or alternate. On the thicker stems, their distribution becomes more irregular, and the polyps extend all around; they often arise obliquely to the surface of the coenenchyme. The coenenchyme and column wall present white granulations, foreign inclusions, which determine the color of the colony; tentacles and oral disk are usually brown in color. Tentacles are short, rounded at their apex, and dicyclic, with 14 to 16 occurring in each cycle. The mouth is rounded or slit-like and the lips are prominent. The capitular ridges are wedge-shaped and acute and vary in number from 14 to 16. The polyps are capable of complete retraction and present mammiform shape; or they may be slightly longer, and flattened or rounded above, a small aperture remains in the middle. The polyps' diameter and height over the coenenchyme is 2 mm.

Recent and other previous records: Curaçao [24,66], Dominica [66], Jamaica [68], Puerto Rico [26], and Tobago [66].

Remarks: Similar to *H. antumbrosus*, *H. tunicans* was locally common at sites around Curaçao and Sint Eustatius where there was consistent strong current, which may be a prerequisite for the host *Dentitheca* hydroid species. Consequently, both Atlantic *Hydrozoanthus* species often co-occur in the same location, at least around Bonaire, Curaçao [24], and Sint Eustatius [21].

Figure 16. In situ images of *Hydrozoanthus tunicans*; (**a**) image M0054611 (not collected) from Sta. EUX04 (Hangover), Sint Eustatius, depth = 16 m, (**b**) specimen MISE JDR150612-93 from Sta. EUX012 [point 117], Sint Eustatius, depth = 14 m, and (**c**) specimen MISE JDR191030-2-1 from Baby Beach (Pretty Rough) [point 29], Bonaire, depth = 25 m. Scale bar in (**c**) = approximately 1 cm.

Suborder Brachycnemina Haddon & Shackleton, 1891 [50]

Family Sphenopidae Hertwig, 1882 [97]

Genus *Palythoa* Lamouroux, 1816 [98]

3.2.14. (a) *Palythoa caribaeorum* (Duchassaing de Fonbressin & Michelotti, 1860) [23] (Figure 17a,c)

Specimens examined (n = 156). **Aruba.** RMNH.COEL.40308 (12°25′04″ N, 69°51′57″ W [point 1], Seru Colorado (Ceru Cora), East Point, intertidal, 2.v.1955, coll. PWH); RMNH.COEL.40307 (12°36′29″ N, 70°03′13″ W [point 4], Malmok, Arashi Beach, depth unknown, 14.viii.1955, coll. PWH). **Bonaire.** RMNH.COEL.40235 (12°05′19″ N, 68°14′00″ W [point 6], 1 km south of Sorobon, shallow water, 13.iv.1973, coll. JCH); RMNH.COEL.40265 (12°12′01″ N, 68°18′38″ W [point 17], ca. 3 km north of Kralendijk, depth unknown, 9.iv.1973, coll. JCH); RMNH.COEL.40299, 40326 (12°08′02″ N, 68°16′55″ W [point 11], De Hoop, south of Kralendijk, intertidal, 10.ix.1948, coll. PWH); RMNH.COEL.40300 (12°09′53″ N, 68°19′24″ W [point 13], West Point, Klein Bonaire, depth unknown, 28.iii.1955, coll. PWH); RMNH.COEL.40309, 40322 (12°07′00″ N, 68°17′43″ W [point 10], north of Point Vierkant, intertidal, 9.ix.1948, coll. PWH); RMNH.COEL.40320 (12°05′59″ N, 68°13′43″ W [point 9], Lac Boca, 1–2 m depth, 1.xi.1948, coll. PWH); MISE JDR191103-2-10 (20 m depth), JDR191103-2-11, JDR191103-2-12, JDR191103-2-13, JDR191103-2-14, JDR191103-2-15, JDR191103-2-16, JDR191103-2-17, JDR191103-2-18 (12°1′36.3″ N, 68°15′4.74″ W [point 23], Red Slave, 5-10 m depth, 3.xi.2019, coll. JDR. **Curaçao.** MISE JDR170609-2-15, JDR170609-2-16, JDR170609-2-17, JDR170609-2-18, JDR170609-2-19, JDR170609-2-20, JDR170609-2-21, JDR170609-2-22, JDR170609-2-23, JDR170609-2-24 (12°06′33″ N, 68°57′15″ W [point 52], Santa Marta, Water Factory, 5–7 m depth, 9.vi.2017, coll. JDR); MISE JDR170609-2-9, JDR170609-2-10, JDR170609-2-11, JDR170609-2-12, JDR170609-2-13, JDR170609-2-14 (12°06′33″ N, 68°57′15″ W [point 52], Santa Marta, Water Factory, 10 m depth, 9.vi.2017, coll. JDR); MISE JDR170611-5-34, JDR170611-5-35, JDR170611-5-36, JDR170611-5-37, JDR170611-5-38 (12°06′27″ N, 68°56′56″ W [point 51], Double Reef, 10–14 m depth, 11.vi.2017, coll. JDR); MISE JDR170611-5-39, JDR170611-5-40, JDR170611-5-41, JDR170611-5-42, JDR170611-5-43 (12°06′27″ N, 68°56′56″ W [point 51], Double Reef, 7–8 m depth, 11.vi.2017, coll. JDR); MISE JDR170612-7-47, JDR170612-7-48, JDR170612-7-49, JDR170612-7-50, JDR170612-7-51, JDR170612-7-52 (12°05′24″ N, 68°54′19″ W [point 50], Marie Pampoen, 3–5 m depth, 12.vi.2017, coll. JDR); MISE JDR170613-10-63 (12°04′05″ N, 68°51′44″ W [point 42], Caracas Bay, Tugboat, 11 m depth, 13.vi.2017, coll. JDR); MISE JDR170613-10-64 (12°04′05″ N, 68°51′44″ W [point 42], Caracas Bay, Tugboat, <1 m depth, 13.vi.2017, coll. JDR); MISE JDR170613-9-57, JDR170613-9-58, JDR170613-9-59 (12°03′59″ N, 68°51′38″ W [point 40], Director's Bay, <1 m depth, 13.vi.2017, coll. JDR); MISE JDR170618-18-84, JDR170618-18-85, JDR170618-18-86, JDR170618-18-87, JDR170618-18-88 (12°16′02″ N, 69°07′43″ W [point 71], south of Santa Martha Bay, 1–2 m depth, 18.vi.2017, coll. JDR); RMNH.COEL.2991 (coordinates and depth unknown, 1920, coll. C.J. van der Horst). **Saba.** RMNH.COEL.40333 (17°36′29.7″ N, 63°15′07.6″ W [point 79], Sta. LUY-153, <1 m depth, 15.vi.1972, coll. Luymes Exp.); RMNH.COEL.40236 (12°03′59″ N, 68°51′04″ W [point 39], Spaanse Water, tidal pools east of Boca, intertidal, 5.ix.1972, coll. JCH); RMNH.COEL.40243 (12°02′00″ N, 68°44′00″ W [point 35], Oostpunt from open sea, intertidal, August 30, 1972, coll. JCH); RMNH.COEL.40252 (17°37′05″ N, 63°15′26″ W [point 81], Sta. LUY-021 between Fort Bay and Ladder Point, depth unknown, 10.iii.1986, coll. JCH); RMNH.COEL.40267 (17°38′35″ N, 63°13′15″ W [point 83], Cove Bay, 1–2 m depth, x.1972 coll. JCH); RMNH.COEL.40284, 40738 (12°14′49″ N, 69°06′25″ W [point 70], Playa Hulu, intertidal, 19.iii.1949, coll. PWH); RMNH.COEL.40295, 40750 (17°37′00″ N, 63°15′11″ W [point 80], west of Fort Bay, intertidal, 6.x.1963, coll. PWH); RMNH.COEL.40302 (12°16′41″ N, 69°38′39″ W [point 72], Boca Pos Spaño, Spaanse Put, depth unknown, 27.ii.1955, coll. PWH); RMNH.COEL.40335 (17°37′41.8″ N, 63°15′30.9″ W [point 82], Sta. LUY-114, 35 m depth, 26.v.1972, coll. Luymes Exp.). **Saba Bank.** RMNH.COEL.40334 (17°20′00″ N, 63°15′00″ W [point 89], Sta. LUY-067, 20 m depth, 16.v.1972, coll. Luymes Exp.); RMNH.COEL.40337 (17°33′00″ N, 63°18′00″ W [point 93], Sta. LUY-084, 17 m depth, 19.v.1972, coll. Luymes Exp.); RMNH.COEL.40345 (17°16′00″ N, 63°21′00″ W [point 87], Sta. LUY-102, 16 m depth, 24.v.1972, coll. Luymes Exp.); RMNH.COEL.40346

(17°29′00″ N, 63°13′00″ W [point 92], Sta. LUY-144, 16 m depth, 14.vi.1972, coll. Luymes Exp.); RMNH.COEL.40748 (17°16′00″ N, 63°33′00″ W [point 88], Sta. LUY-074, 23 m depth, 18.v.1972, coll. Luymes Exp.). **Sint Eustatius**. MISE JDR150610-15, JDR150610-16, JDR150610-20, JDR150610-23, JDR150610-24 (17°27′53.9″ N, 62°59′00.7″ W [point 101], Sta. EUX008, 15–17 m depth, 10.vi.2015, coll. JDR); MISE JDR150611-30, JDR150611-35, JDR150611-36, JDR150611-37, JDR150611-38, JDR150611-39, JDR150611-40, JDR150611-42, JDR150611-43, JDR150611-44, JDR150611-45, JDR150611-46, JDR150611-47, JDR150611-48, JDR150611-49, JDR150611-50, JDR150611-51, JDR150611-52, JDR150611-54, JDR150611-55, JDR150611-56, JDR150611-57, JDR150611-58, JDR150611-59, JDR150611-60, JDR150611-65, JDR150611-66, JDR150611-67, JDR150611-68, JDR150611-69, JDR150611-70, JDR150611-71, JDR150611-72, JDR150611-73, JDR150611-74, JDR150611-75 (17°28′19.2″ N, 62°59′15.6″ W [point 107], Sta. EUX010, 11–13 m depth, 11.vi.2015, coll. JDR); MISE JDR150612-77 (17°30′57.4″ N, 62°59′21.6″ W [point 120], Sta. EUX011, 18 m depth, 12.vi.2015, coll. JDR); MISE JDR150612-88, JDR150612-89, JDR150612-90 (17°30′57.4″ N, 62°59′21.6″ W [point 120], Sta. EUX011, 10 m depth, 12.vi.2015, coll. JDR); MISE JDR150612-95 (17°30′22.6″ N, 63°00′22.0″ W [point 117], Sta. EUX012, 17 m depth, 12.vi.2015, coll. JDR); MISE JDR150613-102, JDR150613-103, JDR150613-104, JDR150613-105, JDR150613-106, JDR150613-107, JDR150613-111, JDR150613-112, JDR150613-113, JDR150613-114, JDR150613-115 (17°28′56.3″ N, 62°59′20.3″ W [point 113], Sta. EUX013, 2–3 m depth, 13.vi.2015, coll. JDR); MISE JDR150614-121, JDR150614-130, JDR150614-131 (17°27′50.9″ N, 62°59′06.8″ W [point 100], Sta. EUX015, 15 m depth, 14.vi.2015, coll. JDR); MISE JDR150615-136 (17°28′05.6″ N, 62°59′30.3″ W [point 104], Sta. EUX016, 18 m depth, 15.vi.2015, coll. JDR); MISE JDR150617-149 (17°28′48.3″ N, 62°59′39.4″ W [point 112], Sta. EUX020, 17 m depth, 17.vi.2015, coll. JDR); MISE JDR150617-156 (17°27′50.4″ N, 62°59′15.0″ W [point 99], Sta. EUX021, 15 m depth, 17.vi.2015, coll. JDR); MISE JDR150619-163, JDR150619-164, JDR150619-167 (17°31′35.7″ N, 62°59′35.3″ W [point 121], Sta. EUX024, Sint Eustatius, 24–29 m depth, June 19, 2015, coll. JDR); MISE JDR150619-172, JDR150619-173, JDR150619-174, JDR150619-175 (17°30′16.4″ N, 62°57′47.8″ W [point 116], Sta. EUX025, 15 m depth, 19.vi.2015, coll. JDR); RMNH.COEL.40230 (17°28′35″ N, 62°59′11″ W [point 111], Sta. Gallows Bay, 1–3 m depth, 30.ix.1972, coll. JCH); RMNH.COEL.40240 (17°28′58″ N, 62°59′23″ W [point 114], Oranjestad Bay, 2–3 m depth, 30.ix.1972, coll. JCH); RMNH.COEL.40311 (17°28′33″ N, 62°59′11″ W [point 110], south of Gallows Bay, 2 m depth, 15.vii.1949, coll. PWH); RMNH.COEL.40336, 40338 (17°28′02.7″ N, 62°58′44.7″ W [point 103], Sta. LUY-121, 10–15 m depth, 9–10.vi.1972, coll. Luymes Exp.). **Sint Maarten**. RMNH.COEL.40247, 40263 (18°00′31″ N, 63°02′46″ W [point 122], Great Bay near Pointe Blanche, <5 m depth, 17.ix.1972, coll. JCH); RMNH.COEL.40321 (18°01′05″ N, 63°02′44″ W [point 123], Great Bay, east side, intertidal, 11.vi.1949, coll. PWH).

Photographic records (*n* = 112). In situ: Specimens MISE JDR150610-15, JDR150610-16, JDR150610-20, JDR150610-23, JDR150610-24, JDR150611-30, JDR150611-35, JDR150611-36, JDR150611-37, JDR150611-38, JDR150611-39, JDR150611-40, JDR150612-77, JDR150612-88, JDR150612-89, JDR150612-90, JDR150612-95, JDR150613-102, JDR150614-121, JDR150614-130, JDR150614-131, JDR150615-136, JDR150617-149, JDR150617-156, JDR150619-163, JDR150619-164, JDR150619-167, JDR150619-172, JDR150619-173, JDR150619-174, JDR150619-175, JDR170609-2-9, JDR170609-2-10, JDR170609-2-11, JDR170609-2-12, JDR170609-2-13, JDR170609-2-14, JDR170609-2-15, JDR170609-2-16, JDR170609-2-17, JDR170609-2-18, JDR170609-2-19, JDR170609-2-20, JDR170609-2-21, JDR170609-2-22, JDR170609-2-23, JDR170609-2-24, JDR170611-5-34, JDR170611-5-35, JDR170611-5-36, JDR170611-5-37, JDR170611-5-38, JDR170611-5-39, JDR170611-5-40, JDR170611-5-41, JDR170611-5-42, JDR170611-5-43, JDR170612-7-47, JDR170612-7-48, JDR170612-7-49, JDR170612-7-50, JDR170612-7-51, JDR170612-7-52, JDR170613-10-63, JDR170613-10-64, JDR170613-9-57, JDR170613-9-58, JDR170613-9-59, JDR191103-2-10, JDR191103-2-11, JDR191103-2-12, JDR191103-2-13, JDR191103-2-14, JDR191103-2-15, JDR191103-2-16, JDR191103-2-17, JDR191103-2-18; Preserved: Specimens RMNH.COEL.2991, 40230, 40235, 40236, 40240, 40243, 40247, 40252, 40263, 40265, 40267, 40284, 40295, 40299, 40300, 40302, 40307, 40308, 40309, 40311, 40320, 40321, 40322, 40326, 40333, 40334, 40335, 40336, 40337, 40338, 40345, 40346, 40738, 40748, 40750.

Description as in Duchassaing de Fonbressin & Michelotti [23]: The living polyps are color yellow citrus, with 30 to 32 tentacles pointed at the end and wider at the bases. When contracted, the polyps do not present a mammiform shape as in other *Palythoa* species; on the contrary, they form a depression.

Recent and other previous records: Ascension [9], Barbados, Belize [99], Brazil [1,100–141], Canary Islands [142], Cape Verde [4,25,143,144], Colombia [145], Costa Rica [146,147], Cuba [74,148,149], Curaçao [24,150], Jamaica [77,148,151–153], Mexico [94,154,155], Panama [78,135,156–158], Puerto Rico [148], Sint Eustatius [22], Saint Helena [5], Saint Thomas [23,148], U.S. Virgin Islands [158], USA (Florida) [159–161], and Venezuela [70,162,163].

Remarks: This is the most widespread zoantharian in shallow waters of the subtropical and tropical Atlantic Ocean [5], as well the most common zoantharian species across the Caribbean, as evidenced by the records listed above. Of interest is that many specimens from older surveys around Sint Eustatius and Curaçao in the 1920s to 1980s are from intertidal depths, while recent surveys in 2015 and 2017 from the same islands found very few colonies in such shallow waters.

Figure 17. (**a**) *Palythoa caribaeorum* specimen MISE JDR150612-89 in situ at Sta. EUX011 [point 120], Sint Eustatius, depth = 10 m, (**b**) *P. caracasiana*, preserved type specimen RMNH.COEL.2990, coordinates and depth unknown, Curaçao, (**c**) preserved *P. caribaeorum* specimen RMNH.COEL.40265 from ca. 3 km north of Kralendijk [point 17], Sint Eustatius, depth = unknown, and (**d**) *P. mammillosa*, preserved type specimen RMNH.COEL.3810 from Aruba, coordinates, depth, date, and collector unknown. Scale bars = approximately 1 cm.

14(b) *Palythoa caracasiana* Pax, 1924 [150] (Figure 17b)

Specimen examined (*n* = 1). **Curaçao.** Holotype, RMNH.COEL.2990 (coordinates and depth unknown, 1920, coll. C.J. van der Horst).

Photographic records (*n* = 1). Preserved: Specimen RMNH.COEL.2990.

Description as in Pax [150]: The coenenchyme and the polyps are dark brown in preserved specimens. The colony tissues are heavily encrusted with pieces of lime and sand grains but remain flexible and present a significant tensile strength. The outline of the colonies is irregularly oval or polygonal. The polyps rise over the coenenchyme, close to each other, and are 3 to 5 mm in diameter. The characteristic feature of *P. caracasiana* are 14 capitular ridges very well demarcated, which are hardly seen in other species of *Palythoa*. The number of mesenteries varies from 28 to 32.

Recent and other previous records: Curaçao [150].

Remarks: This species is almost certainly *P. caribaeorum*, which like many *Palythoa* and *Zoanthus* species, is known to have considerable intraspecific morphological variation in external colony and polyp features (e.g., [164]). For now, as this is a type specimen, we have listed the species as separate from *P. caribaeorum* pending formal revision of these and similar species (see below). As discussed in Burnett et al. [27], it is likely that inadvertent repeated descriptions of the same species from different localities has overly inflated the species counts in both *Palythoa* and *Zoanthus*.

14(c) *Palythoa horstii* Pax, 1924 [150]

Specimens examined (*n* = 1). **Curaçao.** Holotype, RMNH.COEL.2993 (coordinates and depth unknown, 1920, coll. C.J. van der Horst).

Photographic records (*n* = 1). Preserved: Specimen RMNH.COEL.2993.

Description as in Pax [150]: The colony of preserved specimens present a circular outline, with contracted polyps that do not rise above the surface of the coenenchyme, although in some specimens, the polyps rise in the form of small humps. The polyps present 13 to 14 capitular grooves and in average a diameter of 5 mm. The coenenchyme and polyps are heavily encrusted with calcareous granules, and occasionally sponge spicules and foraminiferous skeletons. The number of tentacles is around 28 and mesenteries 30 in dissected specimens; all mesenteries present a distinct basal channel.

Recent and other previous records: None.

Remarks: As with *Palythoa caracasiana* above, this species is almost certainly *P. caribaeorum*.

14(d) *Palythoa mammillosa* (Ellis & Solander, 1786) [165] (Figure 17d)

Specimens examined (*n* = 2). **Aruba.** RMNH.COEL.3810 (coordinates, depth, date, and collector unknown). **Sint Maarten.** RMNH.COEL.2994 (coordinates and collector unknown, ca. 2 m depth, ii.1959).

Photographic records (*n* = 2). Preserved: Specimens RMNH.COEL.2994, 3810.

Description as in Ellis & Solander [165]: The colony is lather-like and spread over the rock surface. Polyps are tightly clustered together, projected over the coenenchyme in a convex mammiform shape, with a hollow in the middle and with a faint start-like appearance. Each polyp has 12 mesenteries and, when expanded, the same number of tentacles.

Recent and other previous records: Bermudas [93], Colombia [145], Guadeloupe [23], Jamaica [23,165–167], Saint Thomas [23], USA (Florida) [159], and Venezuela [162].

Remarks: Although the original description is lacking, if [92] is correct, based on Figure 17d in his work, there is a high probability that this species is the senior synonym of *P. caribaeorum*. Examination of the type species is needed to confirm or refute this idea. For now, the identity of the species remains uncertain, and we have informally placed it within the *caribaeorum-caracasiana-horstii* species grouping.

3.2.15. *Palythoa grandiflora* (Verrill, 1900) [92] (Figure 18)

Specimens examined (*n* = 7). **Aruba.** RMNH.COEL.40312 (12°32′27″ N, 70°03′51″ W [point 3], Bucuti, south point, intertidal, 17.i.1949, coll. PWH). **Bonaire.** MISE JDR191105-2-2 (12°14′46.86″ N, 68°24′49.5″ W [point 24], Playa Frans, 5 m depth, 5.xi.2019, coll. JDR); MISE JDR191106-3-3 (12°6′54.48″ N, 68°17′39.54″ W [point 31], Dolphin Reef, 13 m depth, 6.xi.2019, coll. JDR). **Curaçao.** RMNH.COEL.40241 (12°02′00″ N, 68°44′00″ W [point 35], Oostpunt from open sea, intertidal, 30.viii.1972, coll. JCH). **Sint Eustatius.** RMNH.COEL.40259 (17°30′28″ N, 62°58′51″ W [point 118], west side of Concordia Bay, intertidal, 1.x.1972, coll. JCH); RMNH.COEL.40276 (17°28′35″ N, 62°59′11″ W [point 111], Gallows Bay, 1–6 m depth, 9.iii.1986, coll. JCH). **Sint Maarten.** RMNH.COEL.40232 (18°04′36″ N, 63°00′55″ W [point 125], Baie de l'Embouchure, 1 m depth, x.1972, coll. JCH).

Photographic records (*n* = 7). In situ: MISE JDR191105-2-2, MISE JDR191106-3-3. Preserved: Specimens RMNH.COEL.40232, 40241, 40259, 40276, 40312.

Description as in Verrill [92]: Polyps are large, often spaced by half their length, and form clusters of 12 to 24 polyps of different sizes. When contracted, the polyps form large, rounded mammillae, often higher than wider, strongly sulcated longitudinally, the grooves (~ 26) converging to the central depression of the summit; surface covered with a firm coat of sand. In partial expansion, the summit becomes considerably swollen or turbinated; in full expansion, broad saucer shaped. Tentacles are 52 to 56 in number, short, subequal, with half as many marginal tentacles. The colony is the color buff or light ochre to dark ochre; disks are dull orange or brownish yellow, usually marked with radial lines and specks of white; tentacles are dull orange, often tipped with white; marginal denticles; flake-white. The polyps are 15 to 20 mm in height, 10 to 13 mm in diameter, and the expanded oral disk 14 to 16 mm in diameter.

Recent and other previous records: Bermudas [92], Brazil [1,131], Canary [8], Cuba [149,168], and USA (Florida) [161].

Remarks: Sibling species of Indo-Pacific *P. mutuki* (Haddon & Shackleton, 1891) [169] as shown by molecular data [161]. Recent surveys in both Sint Eustatius in 2015 and Curaçao in 2017 did not find this species, although it was found in Bonaire. Colonies were seen in the Curaçao Sea Aquarium, however reportedly collected from the "Spaanse Water" site on Curaçao (M Jove, Curaçao Sea Aquarium, personal communication).

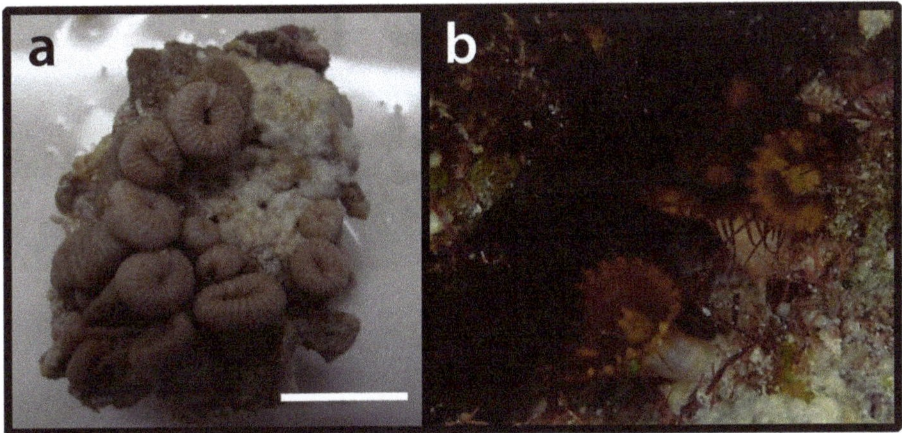

Figure 18. Specimens of *Palythoa grandiflora*; (**a**) preserved specimen RMNH.COEL.40232 from Baie de l'Embouchure [point 125], Sint Maarten, depth = 1 m, and (**b**) specimen MISE JDR191106-3-3 in situ from Dolphin Reef [point 31], Bonaire, depth = 13 m. Scale bar in (**a**) = approximately 1 cm.

3.2.16. *Palythoa grandis* (Verrill, 1900) [92] (Figure 19)

Specimens examined (*n* = 10). **Bonaire.** MISE JDR191101-2-1 (12°15′49.8″ N, 68°24′49.2″ W [point 25], Boka Slaagbaai, 21 m depth, 1.xi.2019, coll. JDR); MISE JGH191107-3-2 (12°10′9.18″ N, 68°18′38.88″ W [point 15], Sampler, Klein Bonaire, 12 m depth, 7.xi.2019, coll. JGH). **Curaçao.** MISE JDR170613-10-62, JDR170619-20-95 (12°04′05″ N, 68°51′44″ W [point 42], Caracas Bay, Tugboat, 11–12 m depth, 13.vi.2017 and 19.vi.2017, coll. JDR); MISE JDR170616-13-77 (12°08′01″ N, 68°59′07″ W [point 62], north of Blue Bay, 11 m depth, 16.vi.2017, coll. JDR) RMNH.COEL.40238 (12°04′29″ N, 68°52′50″ W [point 44], Jan Thiel Bay, 18 m depth, 1.v.1971, coll. J.C. Post); RMNH.COEL.40251 (12°07′20″ N, 68°58′08″ W [point 54], Carmabi House Reef, 64 m depth, 17.viii.1972, coll. J.C. Post). **Sint Eustatius.** MISE JDR150618-161 (17°27′44.5″ N, 62°59′08.1″ W [point 98], Sta. EUX023, 18 m depth, 18.vi.2015, coll. JDR); MISE JDR150619-170 (17°31′35.7″ N, 62°59′35.3″ W [point 121], Sta. EUX024, depth unknown, 19.vi.2015, coll. JDR); MISE JDR150619-178 (17°30′16.4″ N, 62°57′47.8″ W [point 116], Sta. EUX025, 13 m depth, 19.vi.2015, coll. JDR).

Photographic records (*n* = 10). In situ: Specimens MISE JDR170613-10-62, JDR170616-13-77, JDR150618-161, JDR150619-170, JDR150619-178, JDR170619-20-95, JDR191101-2-1, JGH191107-3-2. Preserved: Specimens RMNH.COEL.40238, 40251.

Description as in Verrill [92]: A large species with polyps forming small divergent clusters joined by short stolons, furcate in the base, or sometimes isolated: Walls are thickly encrusted with fine sand. The expanded column is usually clavate, obconic, or long and trumpet-shaped, with the basal part tapered and rather narrow; often two to three times as high as wide. Oral disk broad, cup-shaped, or when fully expanded convex or umbrella shaped, with recurved borders. Tentacles are numerous, about 60 to 66 arranged in two alternating rows, all similar, short, obtuse; outside the tentacles is a circle of marginal papillae, nearly as large as the tentacles and alternating with the outer rows. Sometimes one tentacle (directive) aligned with the longitudinal axis of the mouth is larger and lighter colored than the rest. The color of column is usually pale orange, salmon, or buff, under the coat of white sand; disk is usually orange or orange–brown, sometimes light orange, buff, or ochre–yellow, the color is variable in between the same cluster; the outer part, near the tentacles, is darker than the central, and usually with darker radial lines, sometimes tinged with green; lips are white or orange; the tentacles are similar in color to the disk but usually a shade paler, often darker at base, but the tentacles may be darker than the disk in pale specimens. The largest polyps are 30 to 36 mm in height, and the diameter of the expanded oral disk is 12 to 16 mm.

Recent and other previous records: Barbados [99], Belize [99,170], Bermudas [92,93,171], Canary Islands [142], Costa Rica [146,147], Cuba [149], Curaçao [24], Jamaica [171], Madeira [8], Mexico [94,154,155], Panama [78,157], USA (Florida) [161], and Venezuela [70,162,163].

Remarks: This species is known only from the greater Caribbean, and has no Pacific sibling species, unlike many shallow water zoantharian and *Palythoa* species. Additionally, unlike many *Palythoa* spp., this species seems to be found at deeper areas (e.g., >10 m), and may be a lower-light specialist. One specimen in this study (RMNH.COEL.40251) was found from 64 m, indicating a mesophotic distribution.

Figure 19. In situ images of *Palythoa grandis*; (**a**) specimen MISE JDR170613-10-62 from Caracas Bay, Tugboat [point 42], Curaçao, depth = 11–12 m, and (**b**) specimen MISE JDR150618-161 from Sta. EUX023 [point 98], Sint Eustatius, depth = 18 m. Scale bar in (**a**) = approximately 1 cm.

3.2.17. *Palythoa variabilis* (Duerden, 1898) [76] (Figure 20)

Specimens examined (*n* = 6). **Curaçao.** MISE JDR170617-16-83 (12°11′51″ N, 69°04′46″ W [point 66], Habitat, 37 m depth, 17.vii.2017, coll. JDR); RMNH.COEL.40248 (12°07′45″ N, 68°58′51″ W [point 61], north of Piscadera Bay, 24 m depth, 6.ix.1972, coll. JCH); RMNH.COEL.40736

(12°14′49″ N, 69°06′25″ W [point 70], Playa Hulu, intertidal, 19.iii.1949, coll. PWH). **Sint Eustatius.** RMNH.COEL.40271 (17°30′39″ N, 62°56′52″ W [point 119], boundary of Zeelandia Bay in the north, intertidal, 8.iii.1986, coll. JCH); MISE JDR150621-191 (17°28′12.8″ N, 62°58′58.7″ W [point 105], Sta. EUX029, 3 m depth, 21.vi.2015, coll. JDR); MISE JDR150616-148 (17°28′13.6″ N, 62°59′30.2″ W [point 106], Sta. EUX019, depth unknown, 16.vi.2015, coll. JDR).

Photographic records (*n* = 5). In situ: Specimens MISE JDR150616-148, JDR170617-16-83. Preserved: Specimens RMNH.COEL.40248, 40271, 40736.

Description as in Duerden [76]: Polyps erect, firm, smooth, arising independently from a lamellar coenenchyme, or from the base of one another; or solitary; often cylindrical in retraction; slightly enlarged and flattened distally, or occasionally narrowing and terminating bluntly. The coenenchyme is not well developed and mostly preset only around the base of each polyp. The tissues are heavily incrusted with sand, sponge spicules, diatoms, and test of radiolarians. Capitulum with about 30 ridges and furrows; tentacles acuminate and arranged in two alternating rows of about 30 in each row. Peristome is considerably raised, and the mouth elongated and slit like. In full expansion, the capitulum and disk are much enlarged in proportion to the diameter of the column; the individuals in a colony are closely aggregated, the oral disk is at the same level in all individuals, capitulum margin in contact and by mutual pressure a polygonal outline is formed, leaving no interstices. Thus, a living colony when fully expanded present the appearance of a mosaic work of brown or green depressed disks, with margins of a dark-brown color. The lower part of the columns is a light buff color, and the upper is dark brown; tentacles are usually dark brown, but in some cases olive or green; the disk is either dark brown or bright green, with green radiating lines, and the peristome is the same color as the disk or different but always brown or bright green. In the largest specimen, the length of the column is 5 cm, the diameter 1.2 cm; the average height is 1.5 cm, and diameter 0.7 cm; the diameter of expanded disk is 2.3 cm; tentacles 0.3 cm in length. Given the rigidity of the column wall there is not much contraction in preserved specimens. The mesenteries are usually brachycnemic; in most polyps, 15 perfect mesenteries occur on each side and the same number of imperfect mesenteries.

Recent and other previous records: Brazil [1,100,104,106,109,110,113,114,117,124,130,133–135, 138,172–179], Cape Verde Islands [180], Cuba [74,149,168], Jamaica [76,77,181], and Panama [78,135, 157,182].

Remarks: Widely distributed across the tropical and subtropical Atlantic, and apparently common in some areas [175], this species was only found a handful of times during current surveys. It may be relatively rare in the central Caribbean.

Figure 20. In situ image of *Palythoa variabilis* specimen MISE JDR150616-148 from Sta. EUX019 [point 106], Sint Eustatius, depth = unknown. Scale bar = approximately 5 mm.

3.2.18. *Palythoa* sp. (Figure 21)

Specimens examined (*n* = 1). **Aruba.** RMNH.COEL.40329 (12°32′27″ N, 70°03′51″ W [point 3], Bucuti, south point, intertidal, 17.i.1949, coll. PWH).

Photographic records (*n* = 1). Preserved: Specimen RMNH.COEL.40329.

Remarks: This specimen was in poor condition, making it difficult to identify to species level.

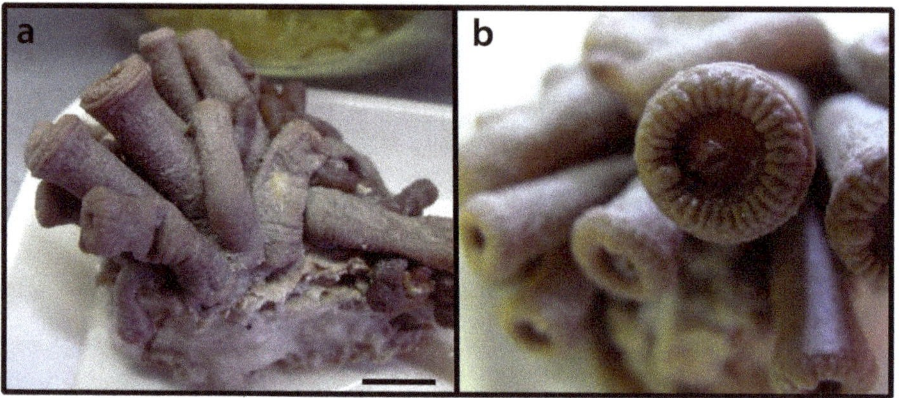

Figure 21. (**a**) and (**b**) images of preserved *Palythoa* sp. specimen RMNH.COEL.40329 from Bucuti, south point [point 3], Aruba, depth = intertidal. Scale bar in (**a**) = approximately 1 cm.

Family Zoanthidae Rafinesque, 1815 [49]
Genus *Zoanthus* Lamarck, 1801 [183]

3.2.19. *Zoanthus pulchellus* (Duchassaing de Fonbressin & Michelotti, 1860) [23] (Figure 22)

Specimens examined (*n* = 15). **Bonaire.** MISE JDR191103-1-4 (12°10′41.1″ N, 68°17′32.34″ W [point 14], Small Wall, 11 m depth, 3.xi.2019, coll. JDR). **Curaçao.** MISE JDR170613-9-56 (12°03′59″ N, 68°51′38″ W [point 40], Director's Bay, 10 m depth, 13.vi.2017, coll. JDR); MISE JDR170616-14-79 (12°08′01″ N, 68°59′07″ W [point 62], south of Blue Bay, 11 m depth, 16.vi.2017, coll. JDR); MISE JDR170621-x-101 (12°04′43″ N, 68°51′37″ W [point 46], Spaanse Water, 1 m depth, 21.vi.2017, coll. JDR); RMNH.COEL.3728 (coordinates and collector unknown, 6 m depth, 21.x.1958); RMNH.COEL.3729 (coordinates and depth unknown, 1920, coll. C.J. van der Horst); RMNH.COEL.40254 (12°07′45″ N, 68°58′51″ W [point 61], north of Piscadera Bay, 24 m depth, 6.ix.1972, coll. P. Creutzberg); RMNH.COEL.40262 (12°07′17″ N, 68°58′09″ W [point 53], Carmabi, Hilton Hotel, 24 m depth, 13.ix.1972, coll. P. Creutzberg & H. de Windt); RMNH.COEL.40327 (12°14′48″ N, 69°06′25″ W [point 69], south of Playa Hulu, intertidal, 2.iv.1949, coll. PWH). **Saba.** RMNH.COEL.40294 (17°37′00″ N, 63°15′11″ W [point 80], west of Fort Bay, intertidal, 6.x.1963, coll. PWH); RMNH.COEL.40340 (17°36′29.7″ N, 63°15′07.6″ W [point 79], Sta. LUY-153, 5–20 m depth, 15.vi.1972, Luymes Exp.). **Sint Eustatius.** MISE JDR150619-171, JDR150619-179, JDR150619-180 (17°30′16.4″ N, 62°57′47.8″ W [point 116], Sta. EUX025, 15–16 m depth, 19.vi.2015, coll. JDR); RMNH.COEL.40257 (17°28′35″ N, 62°59′11″ W [point 111], Sta. LUY-020, Gallows Bay, 1–3 m depth, 30.ix.1972, coll. JCH).

Photographic records (*n* = 14). In situ: Specimens MISE JDR150619-171, JDR150619-179, JDR150619-180, JDR170613-9-56, JDR170616-14-79, JDR191103-1-4. Preserved: Specimens RMNH.COEL.3728, 3729, 40254, 40257, 40262, 40294, 40327, 40340.

Description as in Duchassaing de Fonbressin & Michelotti [23]: Polyps are tightly packed together with oral disk in contact when fully expanded. Tentacles are 60–70 in number, the radio of the oral disk is 5.8 to 7.74 mm in size. The center of the disk is red with green borders and green tentacles.

Recent and other previous records: Bahamas [92], Barbados [99], Brazil [1,105,130,184], Cape Verde [8], Colombia [25], Costa Rica [146], Cuba [74,149,168], Curaçao [24,150], Jamaica [25,76,77,181], Lesser Antilles, Margarita [25], Mexico [94], Panama [78], Puerto Rico [148], Saint Helena [5], Saint Thomas [23,25,76], USA (Florida) [25,161], and Venezuela [70,163].

Remarks: *Zoanthus* spp. are notoriously difficult to identify without either detailed in situ images or histological examinations, and some of the specimens grouped here (and in other *Zoanthus* spp. in this paper) may be incorrectly placed, particularly for older specimens in which no in situ images exist. In particular, recent molecular examinations have indicated that *Zoanthus pulchellus* includes both the species sensu stricto as well as a cryptic species (*Z*. aff. *pulchellus*) that is not yet well delineated morphologically and is different by closely related phylogenetically [161].

This species appears to be, based on recent surveys, distributed in deeper (>10 m) waters in Bonaire, Curaçao, and Sint Eustatius, yet specimens collected from the same islands in the 1940s to 1970s include intertidal specimens. Similar to *P. caribaeorum*, it is not known if this is an artifact from sampling methods (reef-walking vs. SCUBA), due to shallow water bleaching events, or perhaps in this case, specimens of different species inadvertently being put together. More detailed molecular and morphological examinations of both recent and older specimens are needed to examine this, and results may provide important insights into recent climate change.

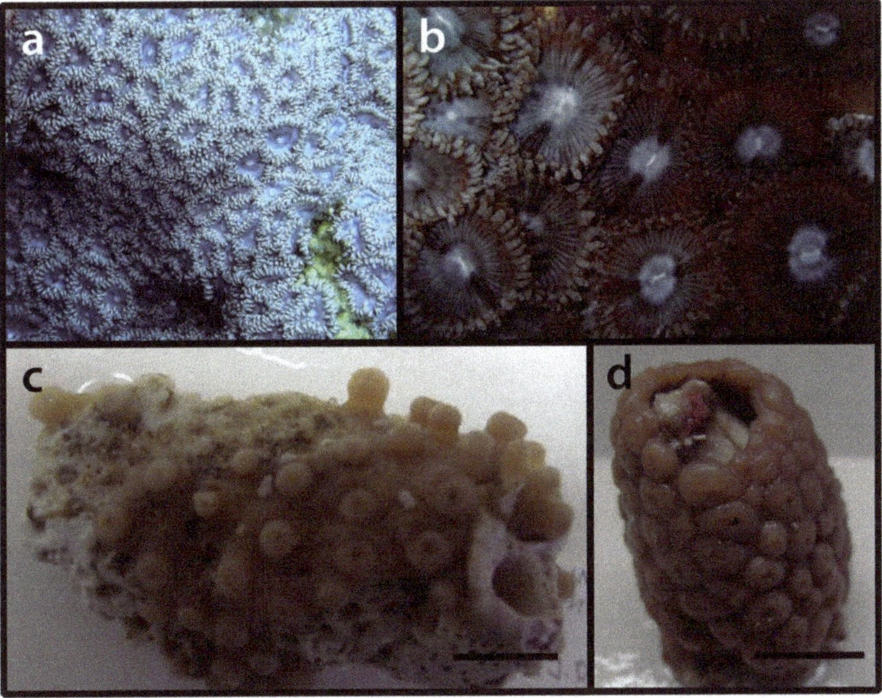

Figure 22. In situ images of *Zoanthus pulchellus*; (**a**) specimen MISE JDR170613-9-56 from Director's Bay [point 40], Curaçao, depth = 10 m, (**b**) specimen MISE JDR191103-1-4 from Small Wall, Bonaire, depth = 11 m, and preserved specimens (**c**) RMNH.COEL.40262 from Carmabi, Hilton Hotel [point 53], Curaçao, depth = 24 m, and (**d**) RMNH.COEL.40257 from Sta. LUY-020, Gallows Bay [point 111], Sint Eustatius, depth = 1–3 m. Scale bars in (**c**) and (**d**) = approximately 1 cm.

3.2.20. *Zoanthus* aff. *pulchellus* (Duchassaing de Fonbressin & Michelotti, 1860) [23]

Specimens examined (*n* = 1). **Curaçao.** MISE JDR170621-x-100 (12°04′43″ N 68°51′37″ W [point 46], Spaanse Water, 1 m depth, 21.vi.2017, coll. JDR).
Photographic Records (*n* = 0).
Description: See Zoanthus pulchellus.
Remarks: This single specimen was morphologically different from *Z. pulchellus* specimens above in being almost completely 'immersae' [25], with polyps barely protruding above a well-developed coenenchyme. However, phylogenetic analyses are needed to confirm the identity of this specimen, particularly given the intraspecific morphological plasticity of *Zoanthus* spp. [164].

3.2.21. *Zoanthus sociatus* (Ellis, 1768) [185] (Figure 23)

Specimens examined (*n* = 33). **Aruba.** RMNH.COEL.3731 (coordinates, depth, date, and collector unknown); RMNH.COEL.40325 (12°30′40″ N, 70°02′18″ W [point 2], Bucuti, north lagoon-side, intertidal, 8.ii.1937, coll. PWH). **Bonaire.** MISE JGH191025-1-3 (12°12′1.74″ N, 68°18′30.72″ W [point 18], Oil Slick, 18 m depth, 25.x.2019, coll. JGH. **Curaçao.** MISE JDR170610-3-27, MISE JDR170610-3-28, MISE JDR170610-3-29 (12°08′21″ N, 68°59′53″ W [point 64], Snake Bay, intertidal, 10.vi.2017, coll. JDR); MISE JDR170613-10-65 (12°04′05″ N, 68°51′44″ W [point 42], Caracas Bay, Tugboat, intertidal, 13.vi.2017, coll. JDR); RMNH.COEL.17819 (12°03′22″ N, 68°50′14″ W [point 37], Fuikbaai, depth unknown, 9.vi.1971, coll. JCH); RMNH.COEL.3730 (coordinates and depth unknown, 1920, coll. C.J. van der Horst); RMNH.COEL.3732 (coordinates and collector unknown, 1–3 m depth, 2.vii.1958). Saba. RMNH.COEL.40749, 40293 (17°37′00″ N, 63°15′11″ W [point 80], west of Fort Bay, intertidal, 6.x.1963, coll. PWH); RMNH.COEL.40261 (12°04′05″ N, 68°51′44″ W [point 42], Caracas Bay, Tugboat, intertidal, 1.vi.1971, coll. JCH); RMNH.COEL.40266 (17°38′35″ N, 63°13′15″ W [point 83], Cove Bay, 1–2 m depth, x.1972, coll. JCH); RMNH.COEL.40285, 12°14′49″ N, 69°06′25″ W [point 70], Playa Hulu, Curaçao, intertidal, March 19, 1949, coll. PWH; RMNH.COEL.40291 (12°03′21″ N, 68°50′05″ W [point 36], Fuikbaai, intertidal, 2.iii.1949, coll. PWH); RMNH.COEL.40297, 12°14′48″ N 69°06′25″ W [point 69], south of Playa Hulu, Curaçao, intertidal, April 2, 1949, coll. PWH; RMNH.COEL.40304 (12°16′41″ N, 69°38′39″ W [point 72], Boca Pos Spaño, Spaanse Put, depth unknown, 27.ii.1955, coll. PWH); RMNH.COEL.40314 (12°19′02″ N, 69°09′09″ W [point 73], Playa Lagun, intertidal, 13.vi.1948, coll. PWH); RMNH.COEL.40323 (coordinates and depth unknown, 1948, coll. PWH). Curaçao. RMNH.COEL.40272 (12°02′00″ N, 68°44′00″ W [point 35], Oostpunt from open sea, intertidal, 30.v.1972, coll. JCH). **Sint Eustatius.** MISE JDR150614-125 (17°27′50.9″ N, 62°59′06.8″ W [point 100], Sta. EUX015, 17 m depth, 14.vi.2015, coll. JDR); MISE JDR150618-160 (17°27′56.6″ N, 63°00′07.2″ W [point 102], Sta. EUX022, 24 m depth, 18.vi.2015, coll. JDR); MISE JDR150619-176 (17°30′16.4″ N, 62°57′47.8″ W [point 116], Sta. EUX025, 14 m depth, 19.vi.2015, coll. JDR); MISE JDR150621-183, MISE JDR150621-184, MISE JDR150621-185 (17°28′12.8″ N, 62°58′58.7″ W [point 105], Sta. EUX029, 3 m depth, 21.vi.2015, coll. JDR); MISE JDR150621-186, JDR150621-188, JDR150621-189, JDR150621-190 (17°28′12.8″ N, 62°58′58.7″ W [point 105], Sta. EUX029, 3–4 m depth, 21.vi.2015, coll. JDR); RMNH.COEL.40275 (17°28′35″ N, 62°59′11″ W [point 111], Gallows Bay, 1–6 m depth, 9.iii.1986, coll. JCH); **Sint Maarten.** RMNH.COEL.40231 (18°05′20″ N, 63°04′55″ W [point 126], Anse des Perres, 1–3 m depth, x.1972, coll. JCH).
Photographic records (*n* = 30). In situ: Specimens MISE JDR150614-125, JDR150618-160, JDR150619-176, JDR150621-183, JDR150621-184, JDR150621-186, JDR150621-188, JDR150621-189, JDR170610-3-27, JDR170610-3-28, JDR170610-3-29, JDR170613-10-65, JGH191025-1-3. Preserved: Specimens RMNH.COEL.3730, 3731, 3732, 40231, 40261, 40266, 40272, 40275, 40285, 40291, 40293, 40297, 40304, 40314, 40323, 40325, 40749.
Description as in Ellis [185]: This compound animal, formed by a tender fleshy substance, consists of many tubular bodies gently elongated towards the upper part and ending in a bulb, similar to a very small onion; on top of which is the mouth, surrounded by one or two rows of tentacles, which when contracted, look like a circle of beads. The lower parts of each tubular body are in communication

with each other through a firm fleshy wrinkled tube, which sticks flat to the rocks. The tubular bodies present multiple sizes rising up irregularly in groups near to one another. The tubular bodies are strongly attached to the coral rock by a fleshy substance with incrusted pieces of shells.

Recent and other previous records: Bahamas [92,186,187], Bermudas [92,93], Brazil [100,105,106, 115,117,121,123,130,131,133–135,138,141,188–191], Cape Verde [8], Colombia [145], Costa Rica [146], Cuba [74,149], Curaçao [150], Dominica [165], Guadeloupe [192], Haiti [25], Jamaica [25,151–153,171, 181,193], Mexico [154,155], most of the West Indies [25], Panama [78,135,156,157], Puerto Rico [148], USA (Florida) [161], and Venezuela [70,162].

Remarks: The first species of zoantharian formally described, this species is widely distributed across the Atlantic [5], and is a sibling species to *Z. sansibaricus* Carlgren, 1900 [194] based on molecular data [161]). It is commonly found forming large colonies in shallow waters, although it was found to 24 m in this study, indicating it may be a depth generalist, similar to *Z. sansibaricus* as seen in Okinawa, Japan [195].

Figure 23. In situ images of *Zoanthus sociatus* (**a**) specimen MISE JDR150621-183 from Sta. EUX029 [point 105], Sint Eustatius, depth = 3 m, (**b**) specimen MISE JDR150621-186 from Sta. EUX029 [point 105], Sint Eustatius, depth = 3–4 m, and preserved specimens (**c**) RMNH.COEL.40314 from Playa Lagun [point 73], Curaçao, depth = intertidal, and (**d**) RMNH.COEL.40291 from Fuikbaai [point 36], Curaçao, depth = intertidal. Scale bar in (**d**) = approximately 1 cm.

3.2.22. *Zoanthus solanderi* Le Sueur, 1817 [192] (Figure 24)

Specimens examined (*n* = 22). **Bonaire.** RMNH.COEL.40301 (12°09′53″ N, 68°19′24″ W [point 13], West Point, Klein Bonaire, depth unknown, 28.iii.1955, coll. PWH); MISE JGH191025-1-2 (12°12′1.74″ N, 68°18′30.72″ W [point 18], Oil Slick, 20 m depth, 25.x.2019, coll. JGH); MISE JDR191031-1-1 (12°9′41.76″ N, 68°17′0.96″ W [point 33], Something Special, 13 m depth, 31.x.2019, coll. JDR); MISE JDR191101-1-1 (12°16′57.42″ N, 68°24′49.32″ W [point 26], Playa Funchi, 20 m depth, 1.xi.2019, coll. JDR); MISE JDR191101-1-2 (12°16′57.42″ N, 68°24′49.32″ W [point 26], Playa Funchi, 20 m depth, 1.xi.2019, coll. JDR); MISE JDR191103-2-5 (12°1′36.3″ N, 68°15′4.74″ W [point 23], Red Slave, 30 m depth, 3.xi.2019, coll. JDR); MISE JDR191104-2-2 (12°4′38.76″ N, 68°16′48″ W [point 28], Invisibles, 10 m depth, 4.xi.2019, coll. JDR); MISE JDR191105-1-3 (12°12′55.38″ N, 68°20′13.08″ W [point 19], Tolo, 6 m depth, 5.xi.2019, coll. JDR). **Curaçao.** MISE JDR170609-2-25 (12°06′33″ N, 68°57′15″ W [point 52], Santa Marta, Water Factory, depth unknown, 9.vi.2017, coll. JDR); MISE JDR170619-20-96 (12°04′05″ N,

68°51′44″ W [point 42], Caracas Bay, Tugboat, 12 m depth, 19.vi.2017, coll. JDR); MISE JDR170620-23-98, JDR170620-23-99 (12°19′45″ N, 69°09′05″ W [point 74], Playa Jeremi, 12–16 m depth, 20.vi.2017, coll. JDR); RMNH.COEL.40303 (12°16′41″ N, 69°38′39″ W [point 72], Boca Pos Spaño, Spaanse Put, depth unknown, 27.ii.1955, coll. PWH); RMNH.COEL.40310, 40737, 40743 (12°14′49″ N, 69°06′25″ W [point 70], Playa Hulu, intertidal, 28.x.1948 and 19.iii.1949, coll. PWH); RMNH.COEL.40330 (coordinates and depth unknown, south side, 13.i.1972, coll. JCH). **Saba.** RMNH.COEL.40292 (17°37′00″ N, 63°15′11″ W [point 80], west of Fort Bay, intertidal, 6.x.1963, coll. PWH). **Sint Eustatius.** MISE JDR150619-169 (17°31′35.7″ N, 62°59′35.3″ W [point 121], Sta. EUX024, 21 m depth, 19.vi.2015, coll. JDR); MISE JDR150619-177 (17°30′16.4″ N,62°57′47.8″ W [point 116], Sta. EUX025, 13 m depth, 19.vi.2015, coll. JDR); MISE JDR150621-187 (17°28′12.8″ N, 62°58′58.7″ W [point 105], Sta. EUX029, 3 m depth, 21.vi.2015, coll. JDR). **Sint Maarten.** RMNH.COEL.40342 (18°04′00″ N, 63°06′00″ W [point 124], Sta. LUY-122, Baie Marigot, <15 m depth, 11.vi.1972, coll. Luymes Exp.).

Photographic records (*n* = 21). In situ: Specimens MISE JDR150619-169, JDR150619-177, JDR150621-187, JDR170609-2-25, JDR170619-20-96, JDR170620-23-98, JGH191025-1-2, JDR191031-1-1, JDR191101-1-1, JDR191101-1-2, JDR191103-2-5, JDR191104-2-2, JDR191105-1-3. Preserved: Specimens RMNH.COEL.40292, 40301, 40303, 40310, 40330, 40342, 40737, 40743.

Description as in Le Sueur [192]: The oral disk of the polyps is of a deep reddish brown color and the peduncle reddish yellow. The tentacles are short and 60 in number. When contracted, the polyp's summit is marked with a deep blue angular spot, and white lines. The polyps are united in groups by the base of their peduncles, in middle of the sand, at the surface of which they raise their discs. The polyps are about 5 cm in length.

Recent and other previous records: Bahamas [92], Brazil [100,112,113,130], Cape Verde [8], Colombia [145], Curaçao [24], Guadeloupe [76], Jamaica [153,181,193,196], Panama [78,135,156,157], Saint Thomas [23,76,192], USA (Florida) [161], and Venezuela [70,162].

Remarks: Although widely distributed, this species is not common, and forms small colonies of <20 cm in diameter. Similar to *P. caribaeorum* and *Z. pulchellus* above, this species was not found in the intertidal zone during recent surveys on Bonaire, Curaçao, and Sint Eustatius despite several specimens from in the intertidal zone in older collections.

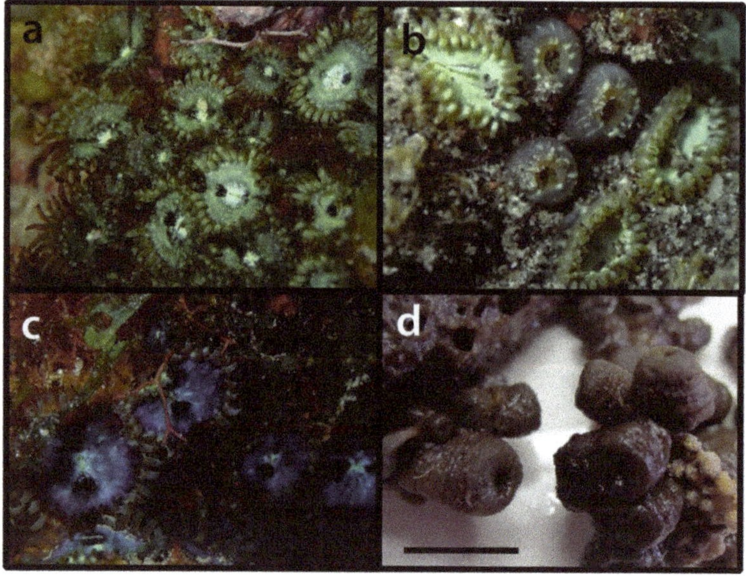

Figure 24. In situ images of *Zoanthus solanderi* (**a**) specimen MISE JDR170619-20-96 from Caracas Bay,

Tugboat [point 42], Curaçao, depth = 12 m, (**b**) specimen MISE JDR150621-187 from Sta. EUX029 [point 105], Sint Eustatius, depth = 3 m, and (**c**) specimen MISE JDR191101-1-1 from Playa Funchi [point 26], Bonaire, depth = 20 m, and (**d**) preserved specimen RMNH.COEL.40292 from west of Fort Bay [point 80], Saba, depth = intertidal. Scale bar in (**d**) = approximately 1 cm.

3.2.23. *Zoanthus* sp. (Figure 25)

Specimens examined (*n* = 10). **Aruba.** RMNH.COEL.40290, 40313, 40328 (12°32′27″ N, 70°03′51″ W [point 3], Bucuti, south point, intertidal, 17.i.1949, coll. PWH). **Bonaire.** RMNH.COEL.40234 (12°13′12″ N, 68°22′41″ W [point 21], southwest coast opposite Lake Goto, shallow water, 11.iv.1978, coll. JCH); RMNH.COEL 40246 (12°05′44″ N, 68°14′14″ W [point 8], Lac Cai, Punta Calbas, depth unknown, 8.iv.1973, coll. JCH). **Curaçao.** RMNH.COEL.40245 (12°03′22″ N, 68°50′14″ W [point 37], Fuikbaai, depth unknown, 9.vi.1971, coll. JCH); RMNH.COEL.40282 (12°03′21″ N, 68°50′05″ W [point 36], Fuikbaai, intertidal, 2.iii.1949, coll. PWH); RMNH.COEL.40296 (12°16′41″ N, 69°38′39″ W [point 72], Boca Pos Spaño, Spaanse Put, depth unknown, 27.ii.1955, coll. PWH); RMNH.COEL.40324 (12°16′02″ N, 69°07′43″ W [point 71], Santa Martha Beach, depth unknown, 29.vii.1955, coll. PWH). **Sint Eustatius.** RMNH.COEL.40242 (17°28′35″ N, 62°59′11″ W [point 111] Gallows Bay, depth unknown, x.1972, coll. JCH).

Photographic records (*n* = 10). Preserved: Specimens RMNH.COEL.40234, 40242, 40245, 40246, 40282, 40290, 40296, 40313, 40324, 40328.

Remarks: These specimens, although some were in good condition, could not be identified to species level, and thus we identify them here to generic level as *Zoanthus* sp. Note, while they are included as one species, material could possibly contain multiple species.

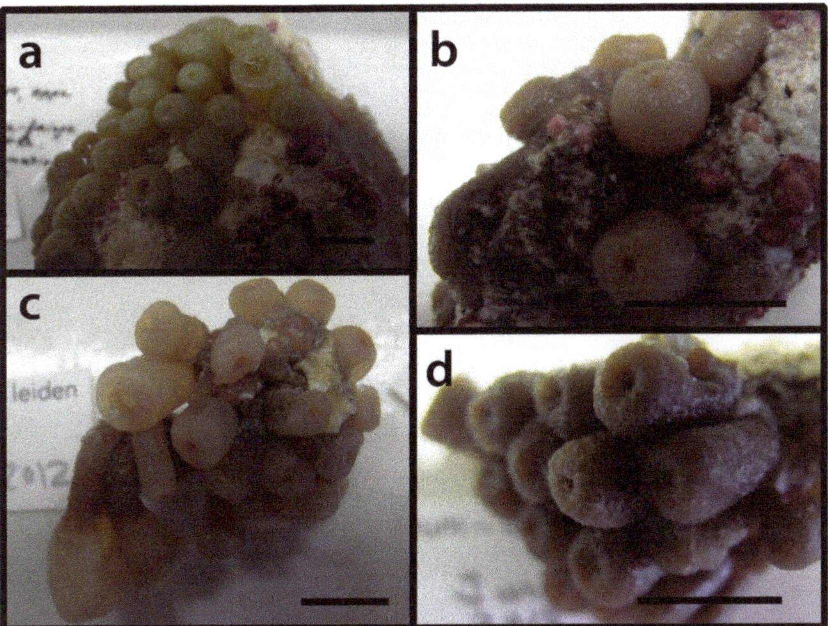

Figure 25. Preserved *Zoanthus* sp. (**a**) specimen RMNH.COEL.40234 from southwest coast opposite Lake Goto [point 21], Bonaire, depth = shallow water, (**b**) specimen RMNH.COEL.40282 from Fuikbaai [point 36], Curaçao, depth = intertidal, (**c**) specimen RMNH.COEL.40245 from Fuikbaai [point 36], Curaçao, depth unknown, and (**d**) specimen RMNH.COEL.40328 from Bucuti, south point [point 3], Aruba, depth = intertidal. Scale bars = approximately 1 cm.

Genus *Isaurus* Gray, 1828 [197]

3.2.24. *Isaurus tuberculatus* Gray, 1828 [197] (Figure 26)

Specimens examined (*n* = 11). **Curaçao.** RMNH.COEL.40244, 40256 (12°04′05″ N, 68°51′44″ W [point 42], Caracas Bay, Tugboat, intertidal, 1.vi.1971, coll. JCH); RMNH.COEL.40260 (12°07′21″ N, 68°58′20″ W [point 55], south of Piscadera Bay, depth unknown, 30.xii.1972 and 13.ix.1972, coll. JCH); RMNH.COEL.40237, 40273 (12°20′27″ N, 68°58′51″ W [point 75], Playa Chikitu (Kleine Knip), 1 m depth, 8.v.1971 and 29.iv.1971, coll. JCH); RMNH.COEL.40283 (12°07′26″ N 68°58′08″ W [point 57], east of Piscadera Bay, depth unknown, 13.ix.1972, coll. JCH), RMNH.COEL.2707, 2708 (coordinates and depth unknown, 1920, coll. C.J. van der Horst). **Sint Eustatius.** MISE JDR150614-129, JDR150614-132 (17°27′50.9″ N, 62°59′06.8″ W [point 100], Sta. EUX015, 15 m depth, 14.vi.2015, coll. JDR); RMNH.COEL.40287 (17°28′32″ N, 62°59′11″ W [point 109], south of Gallows Bay, 2 m depth, 15.vii.1949, coll. PWH).

Photographic records (*n* = 11). In situ: Specimens MISE JDR150614-129, JDR150614-132. Preserved: Specimens RMNH.COEL.2707, 2708, 40237, 40244, 40256, 40260, 40273, 40283, 40287.

Figure 26. *Isaurus tuberculatus* images; (**a**) in situ specimen MISE JDR150614-129 from Sta. EUX015 [point 100], Sint Eustatius, depth = 15 m, and (**b**) preserved specimen RMNH.COEL.40244 from Caracas Bay, Tugboat [point 42], Curaçao, depth = intertidal. Scale bars = approximately 1 cm.

Description as in Gray [197]: Polyps are gregarious, sub-cylindrical, curved, and with longitudinal and transversal grooves and tubercles; tentacles are acute. Polyps are 5.08 cm in length and 1.27 cm in diameter.

Description of *I. gelatinosus* as in Pax [150]: The coenenchyme is poorly developed, and reduced to a lamellar enlargement in the base of the polyps. The shape of the polyps is approximately cylindrical;

some specimens show a clear division between scapus and capitulum, while in others, the differentiation is blurred. The curvatures and tubercles typical of genus *Isaurus* are almost completely absent; even less developed than in var. *microtuberculata*. When tubercles are detectable, they occupied less than the 10% of the total body length and can only be recognized by a careful examination of the material. The diameter of the base of the polyps range from 5 to 18 mm and 11 mm in average; the apical diameter ranges from 4 to 10 mm and 7 mm in average; the polyps length ranges from 20 to 56 mm and 35 mm in average. The tentacles are short and pronounced marginally; the mesenteries were 42 in number and consistent across polyps of different ages.

Recent and other previous records: Ascension [198], Belize [199], Bermudas [93], Brazil [130,138,188,189,200,201], Canary Islands [202], Cape Verde [4], Caribbean [197], Cuba [74,149], Curaçao [150], Guadeloupe [23], Panama [78,157], Saint Helena [5], USA (Florida) [161]. Also reported to the Indo-Pacific [203,204].

Remarks: Specimen RMNH.COEL.2707 was initially described as *Isaurus duchassaingi microtuberculata* Pax, 1924 [150], a subspecies of *Isaurus tuberculatus*. Additionally, RMNH.COEL.2708 is the type specimen for *Isaurus gelatinosus* Pax, 1924 [150]. Both of these species and others were merged into *I. tuberculatus* by Muirhead and Ryland [203].

This species is rare but widely distributed, and in the examined islands, forms small colonies of single or only a few polyps. Currently, based on the revision by Muirhead and Ryland [203], all Atlantic and most Pacific *Isaurus* belong to *I. tuberculatus*, although this is most likely not correct and these are different species, given the isolation of tropical and subtropical zooxanthellate benthic species from each other (discussed in 161).

3.2.25. Zoantharian Species Distribution across the Dutch Caribbean

There are 24 zoantharian species (described and putative) across the Dutch Caribbean in at least nine genera of five families (above and Table 1). Curaçao showed the highest richness of zoantharians (21 species), followed by Bonaire (16), St. Eustatius (15), Saba (7), St. Marten (6), Aruba (5), and Saba Bank (3). In the Southern Caribbean (SC; Aruba, Bonaire and Curaçao), a total of 22 species were observed, while in the Eastern Caribbean (EC; Saba, Saba Bank, St. Eustatius, and St Marten) there were 17 zoantharians. Both SC and EC host species of the genera *Bergia*, *Hydrozoanthus*, *Isaurus*, *Palythoa*, *Parazoanthus*, *Umimayanthus*, and *Zoanthus*. However, *Antipathozoanthus* was only reported from SC, and *Epizoanthus* from EC. There are nine species reported exclusively from the SC and another one from the EC.

Most species occurred in shallow waters (< 200 m depth), with the exception of two species (*Epizoanthus* sp. and Parazoanthidae sp.) found in a deeper zone. Moreover, many species had a wide depth range distribution (Table 1). Six genera (*Bergia*, *Hydrozoanthus*, *Palythoa*, *Parazoanthus*, *Umimayanthus*, *Zoanthus*) occur from the intertidal to the upper mesophotic zone (30–70 m depth), with *Isaurus* extending from the intertidal to 15 m depth. Three colonies of the *Antipathozoanthus* aff. *macaronesicus* were collected at around 10 m depth and *Epizoanthus* sp. was found only in deep waters (980 m).

4. Discussion

Biodiversity loss is one of the gravest threats facing the planet [205] and the Caribbean is under high anthropogenic stress [206–211]. Despite the zoantharian fauna of the Caribbean having been studied for over 250 years, recent works have shown that still much remains to be learned about Caribbean zoantharian diversity, with formal species descriptions ([62]; this study) and indications of other undescribed or unrecognized species ([1,4,161], also as in this study). The Dutch Caribbean is no exception, as our current study describes one new species, *Parazoanthus atlanticus* sp. n., and specimens indicate the presence of other potentially undescribed species (*Antipathozoanthus* aff. *macaronesicus*, *Epizoanthus* sp., *Umimayanthus* sp., *Zoanthus* sp., and Parazoanthidae sp.). Therefore, it is clear that diversity surveys in this region still have much to uncover and are urgently needed.

Furthermore, specimen collection and curating can help provide important baseline data for future analyses. In the current study, we noticed that many zooxanthellate *Palythoa* and *Zoanthus* spp. were collected from intertidal or very shallow waters in surveys until the 1980s, while our more recent surveys between 2015 and 2019 found very few specimens in such shallow waters. While these data alone do not indicate climate change or loss from bleaching, combined with temperature data and information from other taxa, they could be an important part of the picture of how coral reefs have changed over recent history. As our recent expeditions (Curaçao 2017, Bonaire 2019) also surveyed extensively the depth zone between 0 and 3 m, we do not attribute this shift to a change in sampling effort. Additionally, the more recent scuba surveys (2014, 2017, 2019) by trained experts did reveal new depth records for many zoantharian species (Table 1), along with undescribed species, demonstrating the value of having trained taxonomists in the field searching for their specific target taxa.

It is relevant to remark here that many of the present species records are based on historical collection material (RMNH, ZMA), including specimens sampled by C.J. van der Horst from Curaçao (1920), PWH from Aruba (1937, 1955), Bonaire (1948, 1955), Curaçao (1949, 1953, 1955, 1977), Saba (1963), Sint Maarten (1949), and Sint Eustatius (1949), by JCH from Curaçao and Bonaire (1971–1973, 1978, 1986), and the Luymes Expedition to Saba, Saba Bank, Sint Eustatius, and Sint Maarten (1972). Information on some of the sampling localities during which material was collected is well documented in expedition reports [14,15,23]. Examples of Dutch Caribbean zoantharians only known from collections are *Anthipathozoanthus* aff. *macaronesicus* and *Epizoanthus* sp. It is possible that some of these taxa have become, locally, very rare or extinct since they were not observed during intensive fieldwork at Sint Eustatius (2015), Curaçao (2017), and Bonaire (2019). These findings underline the importance of specimen collecting and the continued curation of historical collections [212,213].

Out of the nine families of the order Zoantharia, at least five are recorded to the Dutch Caribbean, covering nine genera and 24 species. Among these, at least five species are endemic to the Caribbean Sea (*Bergia cutressi*, *B. puertoricense*; *Umimayanthus parasiticus*, *Hydrozoanthus antumbrosus*, and *H. tunicans*; [5,26,214]). The Caribbean is the diversity center of reef taxa in the Atlantic Ocean [209,215,216], including Zoantharia [5], and our results provided important insights on the distribution of zoantharians in a regional scale. For example, the SC region had a higher richness than EC (22 and 17 species, respectively; Table 1) and includes the newly described *Parazoanthus atlanticus* sp. n.; a similar pattern was reported by Miloslavich and colleagues [207] for other marine invertebrates across Caribbean regions. Additionally, 17 zoantharian species were distributed across both SC and EC, and some zooxanthellate species were found from intertidal to mesophotic depths, noted by historic and recent surveys. Such results confirm the wide distribution of Zoantharia across localities and depth range [5]. Moreover, two species were reported from deeper than 200 m and are probably unknown to science, highlighting the need of surveys in mesophotic and deep-sea ecosystems [11,217,218].

5. Conclusions

In conclusion, curated specimen collections are an invaluable resource. The Naturalis collections, combined with field surveys, have contributed much to our knowledge of zoantharians in both the Central Indo-Pacific [219] and in this study in the Dutch Caribbean. In addition, photographic records have also been useful during the present research, as previously was demonstrated the South China Sea [220]. For many marine taxa, there is a dearth of information despite their commonality, and basic science as conducted here is a foundation to further studies on different fields and management of species.

Supplementary Materials: The following are available online at http://www.mdpi.com/1424-2818/12/5/190/s1, Figure S1: Phylogenetic reconstructions using the sequences of one nuclear marker (a) ITS-rDNA, and two mitochondrial markers (b) 16S-rDNA, and (c) COI-mtDNA (cytochrome c oxidase I). Table S1: List of specimens examined in this study. Table S2: List of localities recorded in this study.

Author Contributions: Conceptualization, J.D.R. and B.W.H.; field work, J.D.R. and B.W.H.; methodology, all authors; molecular experiments and analyses, J.M., M.E.A.S., H.K., and J.D.R.; resources, J.D.R. and B.W.H.;

data curation, all authors; writing—original draft preparation, J.M.; writing—review and editing, all authors; project administration, funding acquisition, J.M., H.K., J.D.R., and B.W.H. All authors have read and agreed to the published version of the manuscript.

Funding: Research at Bonaire was funded by grants from the WWF Netherlands Biodiversity Fund, the Treub Maatschappij—Society for the Advancement of Research in the Tropics, and the Naturalis program "Nature of the Netherlands". Field work in Sint Eustatius and Curaçao by J.D.R. and visits to the Naturalis collections by J.M. and J.D.R. were supported by Temminck and Martin Fellowships from Naturalis Biodiversity Center. H.K. was partially supported by a Sasagawa Scientific Research grant (2018-5021) from the Japan Science Society.

Acknowledgments: B.W.H. and J.D.R. are grateful to the staff of CARMABI Research Center (Curaçao) and CNSI (Sint Eustatius) for their hospitality and logistical support. BWH is indebted to Adriaan "Dutch" Schrier, Laureen Schenk, and the crews of the *Curasub* and *R/V Chapman*, based at Substation Curaçao, for their generosity and help during the collecting of specimens from deep water. J.D.R. thanks Kelly Latijnhouwers (SECORE International) and Manu Jove (Curaçao Sea Aquarium) for assistance and information on zoantharians in Curaçao. B.W.H. and J.D.R. want to thank STINAPA Bonaire, Dive Friends and Budget Car Rental for logistic support in Bonaire. This publication is Ocean Heritage Foundation/Curaçao Sea Aquarium/Substation Curaçao contribution #OHF/CSA/SC41. J.D.R. thanks all members of the Magnificent 7 for underwater logistics and support and in particular Jaaziel Garcia-Hernandez for sponge identification. Sven Zea (Universidad Nacional de Colombia) is also thanked for help with sponge identification. We are all grateful to two anonymous reviewers for their constructive comments on an earlier draft of the manuscript.

Abbreviations

NA	information not available.
RMNH	Rijksmuseum van Natuurlijke Historie (at Naturalis Biodiversity Center), Leiden, Netherlands
ZMA	Zoological Museum Amsterdam (at Naturalis Biodiversity Center), Leiden, Netherlands
NSMT	National Science Museum, Tsukuba, Ibaraki, Japan
MISE	Molecular Invertebrate Systematics and Ecology Laboratory, University of the Ryukyus, Nishihara, Okinawa, Japan
Por	Porifera collection
Coel	Coelenterata collection
Sta. LUY	Station number Luymes Expedition [47]
Sta. EUX	Station number Marine Biodiversity Expedition to Sint Eustatius, 2015 [48]
coll. BWH	collector: Bert W. Hoeksema
coll. JCH	collector: J.C. (Koos) den Hartog
coll. JDR	collector: James D. Reimer
coll. JGH	collectoer: Jaaziel E. Garcia-Hernandez
coll. PWH	collector: P. Wagenaar Hummelinck

References

1. Santos, M.E.A.; Kitahara, M.V.; Lindner, A.; Reimer, J.D. Overview of the order Zoantharia (Cnidaria: Anthozoa) in Brazil. *Mar. Biodivers.* **2016**, *46*, 547–559. [CrossRef]
2. Fautin, D.G.; Daly, M. Actiniaria, Corallimorpharia, and Zoanthidea (Cnidaria: Anthozoa) of the Gulf of Mexico. In *The Gulf of Mexico—Origins, Waters, and Biota*; Felder, D.L., Camp, D.K., Eds.; Texas A&M University Press: College Station, TX, USA, 2009; pp. 349–357.
3. Humann, P.; DeLoach, N.; Wilk, L. *Reef Creature Identification: Florida, Caribbean, Bahamas*, 3rd ed.; New World Publications: Jacksonville, FL, USA, 2013.
4. Reimer, J.D.; Hirose, M.; Wirtz, P. Zoanthids of the Cape Verde Islands and their symbionts: Previously unexamined diversity in the Northeastern Atlantic. *Contrib. Zool.* **2010**, *79*, 147–163. [CrossRef]
5. Santos, M.E.A.; Wirtz, P.; Montenegro, J.; Kise, H.; López, C.; Brown, J.; Reimer, J.D. Diversity of Saint Helena Island and zoogeography of zoantharians in the Atlantic Ocean: Jigsaw falling into place. *System. Biodivers.* **2019**, *17*, 165–178. [CrossRef]
6. Cairns, S.D.; den Hartog, J.C.; Arneson, C.; Rützler, K. Class Anthozoa (Corals, Anemones). In *Marine Fauna and Flora of Bermuda: A Systematic Guide to the Identification of Marine Organisms*; Sterrer, W., Ed.; Wiley: New York, NY, USA, 1986.

7.   Häussermann, V. Ordnung Zoantharia (=Zoanthinaria, Zoanthidae) (Krustenanemonen). In *Das Mittelmeer—Fauna, Flora, Ökologie. Band II/1: Bestimmungsführer Prokaryota, Protista, Fungi, Plantae, Animalia (bis Nemertea)*; Hofrichter, R., Ed.; Spektrum Akademischer Verlag: Heidelberg, Germany, 2003.

8.   López, C.; Reimer, J.D.; Brito, A.; Simón, D.; Clemente, S.; Hernández, M. Diversity of zoantharian species and their symbionts from the Macaronesian and Cape Verde ecoregions demonstrates their widespread distribution in the Atlantic Ocean. *Coral Reefs* **2019**, *38*, 269–283. [CrossRef]

9.   Reimer, J.D.; Lorion, J.; Irei, Y.; Hoeksema, B.W.; Wirtz, P. Ascension Island shallow-water Zoantharia (*Hexacorallia: Cnidaria*) and their zooxanthellae (*Symbiodinium*). *J. Mar. Biol. Assoc. UK* **2017**, *97*, 695–703. [CrossRef]

10. Carreiro-Silva, M.; Braga-Henriques, A.; Sampaio, I.; de Matos, V.; Porteiro, F.M.; Ocaña, O. *Isozoanthus primnoidus*, a new species of zoanthid (Cnidaria: Zoantharia) associated with the gorgonian *Callogorgia verticillata* (Cnidaria: Alcyonacea). *ICES J. Mar. Sci.* **2011**, *68*, 408–415. [CrossRef]

11. Carreiro-Silva, M.; Ocaña, O.; Stanković, D.; Sampaio, Í.; Porteiro, F.M.; Fabri, M.-C.; Stefanni, S. Zoantharians (*Hexacorallia: Zoantharia*) associated with cold-water corals in the Azores Region: New species and associations in the deep sea. *Front. Mar. Sci.* **2017**, *4*, 88. [CrossRef]

12. Hoeksema, B.W.; Reimer, J.D.; Vonk, R. Editorial: Biodiversity of Caribbean coral reefs (with a focus on the Dutch Caribbean). *Mar. Biodivers.* **2017**, *47*, 1–10. [CrossRef]

13. Van der Horst, C.J. Bijdragen tot de kennis der fauna van Curaçao. Narrative of the voyage and short description of localities. *Bijdr. Dierk.* **1924**, *23*, 1–12, pls. 1–2.

14. Wagenaar Hummelinck, P. Description of new localities. *Stud. Fauna Curaçao Caribb. Isl.* **1953**, *4*, 1–108, pls. 1–8.

15. Wagenaar Hummelinck, P. Marine localities. *Stud. Fauna Curaçao Caribb. Isl.* **1977**, *51*, 1–68, pls. 1–55.

16. Van den Hoek, C.; Cortel-Breeman, A.M.; Wanders, J.B.W. Algal zonation in the fringing coral reef of Curaçao, Netherlands Antilles, in relation to zonation of corals and gorgonians. *Aquat. Bot.* **1975**, *1*, 269–308. [CrossRef]

17. Wanders, J.B.W. The role of benthic algae in the shallow reef of Curaçao (Netherlands Antilles). I: Primary productivity in the coral reef. *Aquat. Bot.* **1976**, *2*, 235–270. [CrossRef]

18. Nagelkerken, I.; Nagelkerken, W.P. Loss of coral cover and biodiversity on shallow *Acropora* and *Millepora* reefs after 31 years on Curaçao, Netherlands Antilles. *Bull. Mar. Sci.* **2004**, *74*, 213–223.

19. Bak, R.P.M. Ecological aspects of the distribution of reef corals in the Netherlands Antilles. *Bijdr. Dierk.* **1975**, *45*, 181–190. [CrossRef]

20. Van Duyl, F.C. *Atlas of the Living Reefs of Curaçao and Bonaire (Netherlands Antilles)*; Foundation for Scientific Research in Surinam and the Netherlands Antilles: Utrecht, The Netherlands, 1985.

21. Reimer, J.D. Zoantharia of St. Eustatius. In *Marine Biodiversity Survey of St. Eustatius, Dutch Caribbean*; Hoeksema, B.W., Ed.; Naturalis and ANEMOON Foundation: Leiden, The Netherlands, 2016; pp. 43–45.

22. Garcia-Hernandez, J.E.; Reimer, J.D.; Hoeksema, B.W. Sponges hosting the Zoantharia-associated crab *Platypodiella spectabilis* at St. Eustatius, Dutch Caribbean. *Coral Reefs* **2016**, *35*, 209. [CrossRef]

23. Duchassaing de Fonbressin, P.; Michelotti, J. *Supplément au mémoire sur les Coralliaires des Antilles. Memorie della Reale Accademia delle Scienze di Torino*; Imprimerie Royale: Turin, Italy, 1860.

24. Reimer, J.D.; Wee, H.B.; García-Hernández, J.E.; Hoeksema, B.W. Zoantharia (Anthozoa: Hexacorallia) abundance and associations with Porifera and Hydrozoa across a depth gradient on the west coast of Curaçao. *System. Biodivers.* **2018**, *16*, 820–830. [CrossRef]

25. Pax, F. Studien an westindischen Actinien. *Zool. Jahrb. Supp.* **1910**, *11*, 157–330.

26. West, D.A. Symbiotic zoanthids (Anthozoa: Cnidaria) of Puerto Rico. *Bull. Mar. Sci.* **1979**, *29*, 253–271.

27. Burnett, W.J.; Benzie, J.A.H.; Beardmore, J.A.; Ryland, J.S. Zoanthids (Anthozoa, Hexacorallia) from the Great Barrier Reef and Torres Strait, Australia: Systematics, evolution and a key to species. *Coral Reefs* **1997**, *16*, 55–68. [CrossRef]

28. Reimer, J.D.; Ono, S.; Fujiwara, Y.; Takishita, K.; Tsukahara, J. Reconsidering *Zoanthus* spp. diversity: Molecular evidence of conspecifity within four previously presumed species. *Zool. Sci.* **2004**, *21*, 517–525. [CrossRef] [PubMed]

29. Swain, T.D.; Wulff, J.L. Diversity and specificity of Caribbean sponge–zoanthid symbioses: A foundation for understanding the adaptive significance of symbioses and generating hypotheses about higher-order systematics. *Biol. J. Linn. Soc.* **2007**, *92*, 695–711. [CrossRef]

30. Rasband, W.S. ImageJ: Image processing and analysis in Java. *ASCL* **2012**, *1*, 6013.

31. England, K.W. Nematocysts of sea anemones (Actiniaria, Ceriantharia and Corallimorpharia: Cnidaria): Nomenclature. *Hydrobiologia* **1991**, *216–217*, 691–697. [CrossRef]

32. Ryland, J.S.; Lancaster, J.E. A review of zoanthid nematocyst types and their population structure. *Hydrobiologia* **2004**, *530–531*, 179–187.

33. Schmidt, H. On evolution in the Anthozoa. In Proceedings of the 2nd International Coral Reef Symposium, Marco Polo (Ship), 22 June–2 July 1973; Cameron, A.M., Campbell, B.M., Cribb, A.B., Endean, R., Jell, J.S., Jones, O.A., Mather, P., Talbot, F.H., Eds.; International Coral Reef Society: Brisbane, Australia, 1974; pp. 533–560.

34. Hidaka, M.; Miyazaki, I.; Yamazato, K. Nematocysts characteristic of the sweeper tentacles of the coral *Galaxea fascicularis* (Linnaeus). *Galaxea* **1987**, *6*, 195–207.

35. Hidaka, M. Use of nematocyst morphology for taxonomy of some related species of scleractinian corals. *Galaxea* **1992**, *11*, 21–28.

36. Folmer, O.; Black, M.; Hoeh, W.; Lutz, R.; Vrijenhoek, R. DNA primers for amplification of mitochondrial cytochrome c oxidase subunit I from diverse metazoan invertebrates. *Mol. Mar. Biol. Biotech.* **1994**, *3*, 294–299.

37. Sinniger, F.; Montoya-Burgos, J.I.; Chevaldonné, P.; Pawlowski, J. Phylogeny of the order Zoantharia (Anthozoa, Hexacorallia) based on the mitochondrial ribosomal genes. *Mar. Biol.* **2005**, *147*, 1121–1128. [CrossRef]

38. Reimer, J.D.; Takishita, K.; Ono, S.; Tsukahara, J.; Maruyama, T. Molecular evidence suggesting interspecific hybridization in *Zoanthus* spp. (Anthozoa: Hexacorallia). *Zool. Sci.* **2007**, *24*, 346–359. [CrossRef]

39. Kearse, M.; Moir, R.; Wilson, A.; Stones-Havas, S.; Cheung, M.; Sturrock, S.; Buxton, S.; Cooper, A.; Markowitz, S.; Duran, C.; et al. Geneious Basic: An integrated and extendable desktop software platform for the organization and analysis of sequence data. *Bioinformatics* **2012**, *28*, 1647–1649. [CrossRef] [PubMed]

40. Katoh, K.; Standley, D.M. MAFFT multiple sequence alignment software version 7: Improvements in performance and usability. *Mol. Biol. Evol.* **2013**, *30*, 772–780. [CrossRef]

41. Edgar, R.C. MUSCLE: Multiple sequence alignment with high accuracy and high throughput. *Nuc. Acids Res.* **2004**, *32*, 1792–1797. [CrossRef]

42. Milne, I.; Lindner, D.; Bayer, M.; Husmeier, D.; McGuire, G.; Marshall, D.F.; Wright, F. TOPALi v2: A rich graphical interface for evolutionary analyses of multiple alignments on HPC clusters and multi-core desktops. *Bioinformatics* **2009**, *25*, 126–127. [CrossRef]

43. Stamatakis, A. RAxML version 8: A tool for phylogenetic analysis and post-analysis of large phylogenies. *Bioinformatics* **2014**, *30*, 1312–1313. [CrossRef] [PubMed]

44. Ronquist, F.R.; Huelsenbeck, J.P. MRBAYES: Bayesian inference of phylogeny. *Bioinformatics* **2003**, *19*, 1572–1574. [CrossRef] [PubMed]

45. Avise, J.C.; Ball, R.M., Jr. Principles of genealogical concordance in species concepts and biological taxonomy. In *Oxford Surveys in Evolutionary Biology*; Futuyama, D., Antonovics, J., Eds.; Oxford University Press: Oxford, UK, 1990; Volume 7, pp. 45–67.

46. Sites, J.W., Jr.; Marshal, J.C. Operational criteria for delimiting species. *Ann. Rev. Ecol. Evol. System.* **2004**, *35*, 199–227. [CrossRef]

47. Van der Land, J. The Saba Bank—A large atoll in the Northeastern Caribbean. *FAO Fish. Rep.* **1977**, *200*, 469–481.

48. Hoeksema, B.W. *Marine Biodiversity Survey of St. Eustatius, Dutch Caribbean, 2015*; Naturalis Biodiversity Center: Leiden, The Netherlands; ANEMOON Foundation: Leiden, The Netherlands, 2016.

49. Rafinesque, C.S. *Analyse de la Nature, ou Tableau de L'univers et des Corps Organisés*; Selbstverl: Palerme, Italy, 1815.

50. Haddon, A.C.; Shackleton, A.M. A revision of the British Actiniae. Part II. The Zoantheae. In Reports on the zoological collections made in the Torres Straits by A.C. Haddon, 1888–1889. *Sci. Trans. Roy. Dublin Soc.* **1891**, *4*, 609–660.

51. Delage, Y.; Hérouard, E. Zoanthidés–Zoanthidae. In *Traité de Zoologie Concrète. Tome II—2me Partie*; Les Coelentérés: Paris, France, 1901.

52. Swain, T.D.; Swain, L.M. Molecular parataxonomy as taxon description: Examples from recently named Zoanthidea (Cnidaria: Anthozoa) with revision based on serial histology of microanatomy. *Zootaxa* **2014**, *3796*, 81–107. [CrossRef]

53. Swain, T.D.; Schellinger, J.L.; Strimaitis, A.M.; Reuter, K.E. Evolution of anthozoan polyp retraction mechanisms: Convergent functional morphology and evolutionary allometry of the marginal musculature in order Zoanthidea (Cnidaria: Anthozoa: Hexacorallia). *BMC Evol. Biol.* **2015**, *15*, 123. [CrossRef] [PubMed]

54. Carlgren, O. Zoantharia. *Dan. Ingolf-Exped.* **1913**, *5*, 1–64.

55. Carlgren, O. Ceriantharia und Zoantharia der deutschen Tiefsee-Expedition. Zoantharia. In *Wissenschaftliche Ergebnisse der Deutschen Tiefsee-Expedition auf dem Dampfer "Valdivia" 1898–1899*; Gustav Fischer: Jena, Germany, 1923; Volume 19, pp. 252–337.

56. Reiswig, H.M.; Dohrmann, M. Three new species of glass sponges (Porifera: Hexactinellida) from the West Indies, and molecular phylogenetics of *Euretidae* and *Auloplacidae* (Sceptrulophora). *Zool. J. Linn. Soc.* **2014**, *171*, 233–253. [CrossRef]

57. Van Soest, R.W.M.; Meesters, E.H.W.G.; Becking, L.E. Deep-water sponges (Porifera) from Bonaire and Klein Curaçao, Southern Caribbean. *Zootaxa* **2014**, *3878*, 401–443. [CrossRef] [PubMed]

58. Sinniger, F.; Reimer, J.D.; Pawlowski, J. The Parazoanthidae (Hexacorallia: Zoantharia) DNA taxonomy: Description of two new genera. *Mar. Biodivers.* **2010**, *40*, 57–70. [CrossRef]

59. Ocaña, O.; Brito, A. A review of Gerardiidae (Anthozoa: Zoantharia) from the Macaronesian islands and the Mediterranean Sea with the description of a new species. *Rev. Acad. Canaria Cien.* **2003**, *15*, 159–189.

60. Kise, H.; Fujii, T.; Masucci, G.D.; Biondi, P.; Reimer, J.D. Three new species and the molecular phylogeny of *Antipathozoanthus* from the Indo-Pacific Ocean (Anthozoa, Hexacorallia, Zoantharia). *ZooKeys* **2017**, *725*, 97–122. [CrossRef]

61. Reimer, J.D.; Fujii, T. Four new species and one new genus of zoanthids (Cnidaria, Hexacorallia) from the Galapagos Islands. *ZooKeys* **2010**, *42*, 1–36. [CrossRef]

62. Swain, T.D. *Isozoanthus antumbrosus*, a new species of zoanthid (Cnidaria: Anthozoa: Zoanthidea) symbiotic with Hydrozoa from the Caribbean, with a key to hydroid and sponge-symbiotic zoanthid species. *Zootaxa* **2009**, *2051*, 41–48. [CrossRef]

63. Duerden, J.E. West Indian sponge incrusting actinians. *Bull. Am. Mus. Nat. Hist.* **1903**, *19*, 495–503, pls. 44–47.

64. Lewis, T.B.; Finelli, C.M. Epizoic zoanthids reduce pumping in two Caribbean vase sponges. *Coral Reefs* **2015**, *34*, 291–300. [CrossRef]

65. Crocker, L.A.; Reiswig, H.M. Host specificity in sponge-encrusting Zoanthidea (Anthozoa: Zoantharia) of Barbados, West Indies. *Mar. Biol.* **1981**, *65*, 231–236. [CrossRef]

66. Swain, T.D. Phylogeny-based species delimitations and the evolution of host associations in symbiotic zoanthids (Anthozoa, Zoanthidea) of the wider Caribbean region. *Zool. J. Linn. Soc.* **2009**, *156*, 223–238. [CrossRef]

67. Montenegro, J.; Acosta, A. Habitat preference of Zoantharia genera depends on host sponge morphology. *Universitas Scientiarum* **2010**, *15*, 110–121. [CrossRef]

68. Fontaine, A. The colonial sea-anemones of Jamaica. *Nat. Hist. Notes Nat. Hist. Soc. Jamaica* **1954**, *66*, 107–109.

69. Sammarco, P.W.; Porter, S.A.; Genazzio, M.; Sinclair, J. Success in competition for space in two invasive coral species in the western Atlantic—*Tubastraea micranthus* and *T. coccinea*. *PLoS ONE* **2015**, *10*, e0144581. [CrossRef] [PubMed]

70. González-Muñoz, R.; Simões, N.; Guerra-Castro, E.J.; Hernández-Ortíz, C.; Carrasquel, G.; Mendez, E.; Lira, C.; Rada, M.; Hernández, I.; Pauls, S.M. Sea anemones (Cnidaria: Actiniaria, Corallimorpharia, Ceriantharia, Zoanthidea) from marine shallow-water environments in Venezuela: New records and an updated inventory. *Mar. Biodivers. Rec.* **2016**, *9*, 18. [CrossRef]

71. Montenegro, J.; Low, M.E.Y.; Reimer, J.D. The resurrection of the genus *Bergia* (Anthozoa, Zoantharia, Parazoanthidae). *Syst. Biodivers.* **2016**, *14*, 63–73. [CrossRef]

72. Gray, J.E. Notes on Zoanthinae, with the descriptions of some new genera. *Proc. Zool. Soc. Lond.* **1867**, *1*, 233–240.

73. Irving, R.A. A preliminary investigation of the sublittoral habitats and communities of Ascension Island. *Prog. Underw. Sci.* **1989**, *13*, 65–78.

74. Diez, Y.L.; Campos-Castro, A. Soft corals (Anthozoa: Corallimorpharia, Actinaria and Zoantharia) from southeastern of Cuba, and its distribution in Marine Protected Areas. *Rev. Investig. Mar.* **2016**, *36*, 80–93.

75. Williams, E.H., Jr.; Clavijo, I.; Kimmel, J.J.; Colin, P.L.; Carela, C.D.; Bardales, A.T.; Armstrong, R.H.; Williams, L.B.; Boulon, R.H.; Garcia, J.R. A checklist of marine plants and animals of the south coast of the Dominican Republic. *Carib. J. Sci.* **1983**, *19*, 39–53.

76. Duerden, J.E. Jamaican Actiniaria. Part, I.-Zoantheae. *Sci. Trans. Roy. Dublin Soc.* **1898**, *6*, 329–385.

77. Duerden, J.E. Jamaican Actiniaria. Part II. Stichodactylinae and Zoantheae. *Sci. Trans. Roy. Dublin Soc.* **1900**, *7*, 133–200.

78. Cubit, J.D.; Williams, S. The invertebrates of Galeta Reef (Caribbean Panama): A species list and bibliography. *Atoll Res. Bull.* **1983**, *269*, 1–45. [CrossRef]

79. Norman, A.M. Shetland final dredging report—Part II. On the Crustacea, Tunicata, Polyzoa, Echinodermata, Actinozoa, Hydrozoa, and Porifera. In Proceedings of the Thirty-Eighth Meeting of the British Association for the Advancement of Science, Norwich, UK, 19–26 August 1868; John Murray: London, UK, 1869; pp. 247–336.

80. Carlgren, O. Actiniaria and Zoantharia. In *Further Zoological Results of the Swedish Antartic Expedition 1901-1903 under the Direction of Dr. Otto Nordnskjold*, 1st ed.; Odhner, T., Ed.; Norstedt and Soner: Stockholm, Sweden, 1927; pp. 93–95.

81. Pax, F. *Aktinien der Aru-Inseln*; Senckenbergischen Naturforschenden Gesellschaft: Frankfurt, Germany, 1911; pp. 297–304.

82. McMurrich, J.P. The Actiniae of the Plate Collection. *Zool. Jahrb. Supp.* **1904**, *6*, 215–306.

83. Carlgren, O. Actiniaria und Zoantharia von Juan Fernandez und der Osterinsel. In *The Natural History of Juan Fernandez and Easter Island*; Skottsberg, C., Ed.; Almquist & Wiksells Boktryckeri: Upsalla, Sweden, 1922; Volume 3, Pt 2, pp. 145–160.

84. Cutress, C.E. Chapter 7, Corallimorpharia, Actiniaria and Zoanthidea. In *Memoirs of the National Museum of Victoria 32*; Gill, E.D., Ed.; National Museum of Victoria: Melbourne, Australia, 1971; pp. 89–90.

85. Schmidt, O. *Die Spongien des Adriatischen Meeres*, 1st ed.; Wilhelm Engelmann: Leipzig, Germany, 1862; pp. 1–88.

86. Duerden, J.E. A new species of *Parazoanthus*. In *Records Albany Museum 2*; Albany Museum: Grahamstown, South Africa, 1907; p. 80.

87. Montenegro, J.; Sinniger, F.; Reimer, J.D. Unexpected diversity and new species in the sponge-Parazoanthidae association in southern Japan. *Mol. Phylogenet. Evol.* **2015**, *89*, 73–90. [CrossRef]

88. Kobuk, D.R.; Van Soest, R.W. Cavity-dwelling sponges in a southern Caribbean coral reef and their paleontological implications. *Bull. Mar. Sci.* **1989**, *44*, 1207–1235.

89. Ryland, J.S. Reproduction in Zoanthidea (Anthozoa: Hexacorallia). *Invertebr. Reprod. Dev.* **1997**, *31*, 177–188. [CrossRef]

90. Swain, T.D. Context-dependent effects of symbiosis: Zoanthidea colonization generally improves Demospongiae condition in native habitats. *Mar. Biol.* **2012**, *159*, 1429–1438. [CrossRef]

91. Ryland, J.S.; Westphalen, D. The reproductive biology of *Parazoanthus parasiticus* (Hexacorallia: Zoanthidea) in Bermuda. *Hydrobiologia* **2004**, *530/531*, 411–419. [CrossRef]

92. Verrill, A.E. Additions to the Anthozoa and Hydrozoa of the Bermudas. *Trans. Conn. Acad. Arts Sci.* **1900**, *10*, 551–572. [CrossRef]

93. Verrill, A.E. The Bermuda Islands: Part, V. An account of the coral reefs (characteristic life of the Bermuda coral reefs). *Trans. Conn. Acad. Arts Sci.* **1907**, *12*, 280–296.

94. De la Cruz-Francisco, V.; González-González, M.; Morales-Quijano, I. Inventario taxonómico de Hydrozoa (Orden: Anthoathecata) y Anthozoa (Subclases: Hexacorallia y Octocorallia) del Arrecife Enmedio, Sistema Arrecifal Lobos-Tuxpan. *CICIMAR Oceánides* **2016**, *31*, 23–34.

95. Hill, A.; Wagner, A.; Hill, M. Hox and paraHox genes from the anthozoan *Parazoanthus parasiticus*. *Mol. Phylogenet. Evol.* **2003**, *28*, 529–535. [CrossRef]

96. Hill, M.S. Sponges harbor genetically identical populations of the zoanthid *Parazoanthus parasiticus*. *Bull. Mar. Sci.* **1998**, *63*, 513–521.

97. Hertwig, R. Report on the Actiniaria dredged by H.M.S. Challenger during the years 1873–1876. In *Reports on the Scientific Results of the Exploring Voyage of H.M.S. Challenger during the Years 1873–1876*; Neill: Edinburgh, UK, 1882; pp. 1–134.

98. Lamouroux, J.V. *Histoire des Polypiers Coralligenes Flexibles, Vulgairement Nommés Zoophytes*; F. Poisson: Caen, France, 1816.

99. Finney, J.C.; Pettay, D.T.; Sampayo, E.M.; Warner, M.E.; Oxenford, H.A.; LaJeunesse, T.C. The relative significance of host–habitat, depth, and geography on the ecology, endemism, and speciation of coral endosymbionts in the genus *Symbiodinium*. *Microb. Ecol.* **2010**, *60*, 250–263. [CrossRef]

100. Acosta, A. Disease in zoanthids: Dynamics in space and time. *Hydrobiologia* **2001**, *460*, 113–130. [CrossRef]

101. Acosta, A.; Sammarco, P.W.; Duarte, L.F. Asexual reproduction in a zoanthid by fragmentation: The role of exogenous factors. *Bull. Mar. Sci.* **2001**, *68*, 363–381.

102. Acosta, A.; Sammarco, P.W.; Duarte, L.F. New fission processes in the zoanthid *Palythoa caribaeorum*: Description and quantitative aspects. *Bull. Mar. Sci.* **2005**, *76*, 1–26.

103. Acosta, A.; González, A.M. Fission in the Zoantharia *Palythoa caribaeorum* (Duchassaing and Michelotii, 1860) populations: A latitudinal comparison. *Boletín de Investig. Mar. Costeras INVEMAR* **2007**, *36*, 151–165. [CrossRef]

104. Almeida, J.G.L.; Maia, A.I.; Wilke, D.; Silveira, E.R.; Braz-Filho, R.; La Clair, J.J.; Costa-Lotufo, L.; Pessoa, O.D.L. Palyosulfonoceramides A and B: Unique sulfonylated ceramides from the Brazilian zoanthids *Palythoa caribaeorum* and *Protopalyhtoa variabilis*. *Mar. Drugs* **2012**, *10*, 2846–2860. [CrossRef] [PubMed]

105. Amaral, F.D.; Hudson, M.M.; da Silveira, F.L.; Migotto, A.E.; Pinto, S.M.; Longo, L. Cnidarians of Saint Peter and St. Paul Archipelago, Northeast Brazil. In Proceedings of the 9th International Coral Reef Symposium, Bali, Indonesia, 23–27 October 2000; Moosa, M.K., Soemodihardjo, S., Soegiarto, A., Romimohtarto, K., Nontji, A., Soekarno, S., Eds.; International Coral Reef Society: Bali, Indonesia, 2000; pp. 567–572.

106. Amaral, F.M.D.; Ramos, C.A.C.; Leão, Z.; Kikuchi, R.K.P.; Lima, K.K.M.; Longo, L.L.; Cordeiro, R.T.S.; Lira, S.M.A.; Vasconcelos, S.L. Checklist and morphometry of benthic cnidarians from the Fernando de Noronha Archipelago, Brazil. *Cah. Biol. Mar.* **2009**, *50*, 277–290.

107. Azevedo, C.A.A.; Carneiro, M.A.A.; Oliveira, S.R.; Marinho-Soriano, E. Macrolgae as an indicator of the environmental health of the Pirangi reefs, Rio Grande do Norte, Brazil. *Rev. Brasil. Farmacog.* **2011**, *21*, 323–328. [CrossRef]

108. Barreira, C.; Echeverría, C.A.; de Oliveira Pires, D.; Fonseca, C.G. Distribuição do bentos (Cnidaria e Echinodermata) em costões rochosos da Baía de Ilha Grande, Rio de Janeiro, Brasil. *Oecol. Brasil* **1999**, *7*, 179–193.

109. Boscolo, H.K.; Silveira, F.L. Reproductive biology of *Palythoa caribaeorum* and *Protopalythoa variabilis* (Cnidaria, Anthozoa, Zoanthidea) from the southeastern coast of Brazil. *Braz. J. Biol.* **2005**, *65*, 29–41. [CrossRef]

110. Bouzon, J.L.; Brandini, F.P.; Rocha, R.M. Biodiversity of sessile fauna on rocky shores of coastal islands in Santa Catarina, Southern Brazil. *Mar. Sci.* **2012**, *2*, 39–47. [CrossRef]

111. Castro, C.B.; Segal, B.; Negrão, F.; Calderon, E.N. Four-year monthly sediment deposition on turbid southwestern Atlantic coral reefs, with a comparison of benthic assemblages. *Braz. J. Oceanog.* **2012**, *60*, 49–63. [CrossRef]

112. Chimetto, L.A.; Brocchi, M.; Thompson, C.C.; Martins, R.C.R.; Ramos, H.R.; Thompson, F.L. Vibrios dominate as culturable nitrogen-fixing bacteria of the Brazilian coral *Mussismilia hispida*. *System. App. Microb.* **2008**, *31*, 312–319. [CrossRef]

113. Chimetto, L.A.; Cleenwerck, I.; Thompson, C.C.; Brocchi, M.; Willems, A.; De Vos, P.; Thompson, F.L. *Photobacterium jeanii* sp. nov., isolated from corals and zoanthids. *Int. J. System. Evol. Microb.* **2010**, *60*, 2843–2848. [CrossRef]

114. Chimetto, L.A.; Cleenwerck, I.; Moreira, A.P.B.; Brocchi, M.; Willems, A.; De Vos, P.; Thompson, F.L. *Vibrio variabilis* sp. nov. and *Vibrio maritimus* sp. nov., isolated from *Palythoa caribaeorum*. *Int. J. System. Evol. Microb.* **2011**, *61*, 3009–3015. [CrossRef] [PubMed]

115. Correia, M.D.; Sovierzoski, H.H. Macrobenthic diversity reaction to human impacts on Maceió coral reefs, Alagoas, Brazil. In Proceedings of the 11th International Coral Reef Symposium, Fort Lauderdale, FL, USA, 7–11 July 2008; Cunning, J.R., Thurmond, J.E., Smith, G.W., Weil, E., Ritchie, K.B., Eds.; International Coral Reef Society: Davie, FL, USA, 2008; pp. 1083–1087.

116. Costa, D.L.; Gomes, P.B.; Santos, A.M.; Valença, N.S.; Vieira, N.A.; Pérez, C.D. Morphological plasticity in the reef zoanthid *Palythoa caribaeorum* as an adaptive strategy. *Ann. Zool. Fenn.* **2011**, *48*, 349–358. [CrossRef]

117. Da Silveira, F.L.; Morandini, A.C. Checklist dos Cnidaria do estado de São Paulo, Brasil. *Biota Neotropica* **2011**, *11* (Suppl. S1), 445–454. [CrossRef]

118. De Andrade Melo, L.F.; da Camara, C.A.G.; de Albuquerque Modesto, J.C.; Pérez, C.D. Toxicity against *Artemia salina* of the zoanthid *Palythoa caribaeorum* (Cnidaria: Anthozoa) used in folk medicine on the coast of Pernambuco, Brazil. *Biotemas* **2012**, *25*, 145–151.

119. de Barros, M.M.L.; Castro, C.B.; Pires, D.O.; Segal, B. Coexistence of reef organisms in the Abrolhos Archipelago, Brazil. *Rev. Biol. Trop.* **2000**, *48*, 741–747.

120. De Santana, E.F.C.; Alves, A.L.; Santos, A.D.M.; Maria Da Gloria, G.S.; Perez, C.D.; Gomes, P.B. Trophic ecology of the zoanthid *Palythoa caribaeorum* (Cnidaria: Anthozoa) on tropical reefs. *J. Mar. Biol. Assoc. UK* **2015**, *95*, 301–309. [CrossRef]

121. Echeverría, C.; Pires, D.; Medeiros, M.; Castro, C. Cnidarians of the Atol das Rocas, Brazil. In Proceedings of the 8th International Coral Reef Symposium, Panama, Panama, 24–29 June 1996; Lessios, H.A., Macintyre, I.G., Eds.; International Coral Reef Society: Panama, Panama, 1997; pp. 443–446.

122. Francini-Filho, R.B.; Ferreira, C.M.; Coni, E.O.C.; De Moura, R.L.; Kaufman, L. Foraging activity of roving herbivorous reef fish (Acanthuridae and Scaridae) in eastern Brazil: Influence of resource availability and interference competition. *J. Mar. Biol. Assoc. UK* **2010**, *90*, 481–492. [CrossRef]

123. Francini-Filho, R.B.; de Moura, R.L. Predation on the toxic zoanthid *Palythoa caribaeorum* by reef fishes in the Abrolhos Bank, eastern Brazil. *Braz. J. Oceanog.* **2010**, *58*, 77–79. [CrossRef]

124. Kelecom, A.; Solé-Cava, A.M. Comparative study of zoanthid sterols the genus *Palythoa* (Hexacorallia, Zoanthidea). *Comp. Biochem. Physiol. Part B Comp. Biochem.* **1982**, *72*, 677–682. [CrossRef]

125. Longo, G.O.; Krajewski, J.P.; Segal, B.; Floeter, S.R. First record of predation on reproductive *Palythoa caribaeorum* (Anthozoa: Sphenopidae): Insights on the trade-off between chemical defences and nutritional value. *Mar. Biodivers. Rec.* **2012**, *5*, 1–3. [CrossRef]

126. MacCord, F.S.; Duarte, L.F.L. Dispersion in populations of *Tropiometra carinata* (Crinoidea: Comatulida) in the Sao Sebastiao Channel, Sao Paulo State, Brazil. *Est. Coast. Shelf Sci.* **2002**, *54*, 219–225. [CrossRef]

127. Martinez, A.S.; Mendes, L.F.; Leite, T.S. Spatial distribution of epibenthic molluscs on a sandstone reef in the northeast of Brazil. *Braz. J. Biol.* **2012**, *72*, 287–298. [CrossRef] [PubMed]

128. Mendonça-Neto, J.P.; Ferreira, C.E.L.; Chaves, L.C.T.; Pereira, R.C. Influence of *Palythoa caribaeorum* (Anthozoa, Cnidaria) zonation on site-attached reef fishes. *An. Acad. Bras. Ciênc.* **2008**, *80*, 495–513. [CrossRef] [PubMed]

129. Mendonça-Neto, J.P.; da Gama, B.A.P. The native *Palythoa caribaeorum* overgrows on invasive species in the intertidal zone. *Coral Reefs* **2008**, *28*, 497. [CrossRef]

130. Migotto, A.E.; Silveira, F.L.; Schlenz, E.; Castro, C.B.; Marques, A.C. *Lista dos Cnidaria registrados na costa Brasileira. Invertebrados marinhos registrados no litoral Brasileiro*; University of São Paulo: São Paulo, Brazil, 1998; pp. 1–59.

131. Oigman-Pszczol, S.S.; Figueiredo, M.A.d.O.; Creed, J.C. Distribution of benthic communities on the tropical rocky subtidal of Armação dos Búzios, southeastern Brazil. *Mar. Ecol.* **2004**, *25*, 173–190. [CrossRef]

132. Pérez, C.D.; Vila-Nova, D.A.; Santos, A.M. Associated community with the zoanthid *Palythoa caribaeorum* (Duchassaing & Michelotti, 1860) (Cnidaria, Anthozoa) from littoral of Pernambuco, Brazil. *Hydrobiologia* **2005**, *548*, 207–215. [CrossRef]

133. Rabelo, E.F.d.O.; Soares, M.; Matthews-Cascon, H. Competitive interactions among zoanthids (Cnidaria: Zoanthidae) in an intertidal zone of Northeastern Brazil. *Braz. J. Oceanog.* **2013**, *61*, 35–42. [CrossRef]

134. Rabelo, E.F.; Rocha, L.L.; Colares, G.B.; Bomfim, T.A.; Nogueira, V.L.R.; Katzenberger, M.; Matthews-Cascon, H.; Melo, V.M.M. *Symbiodinium* diversity associated with zoanthids (Cnidaria: Hexacorallia) in Northeastern Brazil. *Symbiosis* **2015**, *64*, 105–113. [CrossRef]

135. Sebens, K.P. Autotrophic and heterotrophic nutrition of coral reef zoanthids. In Proceedings of the 3rd International Coral Reef Symposium, Miami, FL, USA, May 1977; Taylor, D.L., Ed.; International Coral Reef Society: Miami, FL, USA, 1977; pp. 397–406.

136. Segal, B.; Castro, C.B. Coral community structure and sedimentation at different distances from the coast of the Abrolhos Bank, Brazil. *Braz. J. Oceanog.* **2011**, *59*, 119–129. [CrossRef]

137. Soares, C.L.S.; Pérez, C.D.; Maia, M.B.S.; Silva, R.S.; Melo, L.F.A. Avaliação da atividade antiinflamatória e analgésica do extrato bruto hidroalcoólico do zoantídeo *Palythoa caribaeorum* (Duchassaing & Michelotti, 1860). *Braz. J. Pharmacog.* **2006**, *16*, 463–468.

138. Soares, M.O.; Rabelo, E.F.; Mathews-Cascon, H. Intertidal anthozoans from the coast of Ceará, Brazil. *Rev. Bras. Biociênc.* **2011**, *9*, 437–443.

139. Souza, D.S.L.; Grossi-de-Sa, M.M.F.; Silva, L.P.; Franco, O.L.; Gomes-Junior, J.E.; Oliveira, G.R.; Rocha, T.L.; Magalhaes, C.P.; Marra, B.M.; Grossi-de-Sa, M.M.F. Identification of a novel β-N-acetylhexosaminidase (Pcb-NAHA1) from marine zoanthid *Palythoa caribaeorum* (Cnidaria, Anthozoa, Zoanthidea). *Prot. Express. Purif.* **2008**, *58*, 61–69. [CrossRef]

140. Stampar, S.N.; da Silva, P.F.; Osmar Jr, J.L. Predation on the zoanthid *Palythoa caribaeorum* (Anthozoa, Cnidaria) by a Hawksbill turtle (*Eretmochelys imbricata*) in Southeastern Brazil. *Mar. Turtle Newsl.* **2007**, *1*, 3–5.

141. Villaça, R.; Pitombo, F.B. Benthic communities of shallow-water reefs of Abrolhos, Brazil. *Rev. Bras. Oceanogr.* **1997**, *45*, 35–43. [CrossRef]

142. Barquin-Diez, J.; Gonzalez-Lorenzo, G.; Martin-Garcia, L.; Candelaria Gil-Rodriguez, M.; Brito-Hernandez, A. Spatial distribution of benthic subtidal communities of shallow waters of the Canary Islands. I: Soft bottom communities of Tenerife coast. *Vieraea* **2005**, *33*, 435–448.

143. Morri, C.; Bianchi, C.N. Cnidarian zonation at Ilha do Sal (Arquipelago de Cabo Verde). *Beitr. Paläont.* **1995**, *20*, 41–49.

144. Morri, C.; Cattaeno-Vietti, R.; Sartoni, G.; Bianchi, C.N. Shallow epibenthic communities of Ilha do Sal (Cape Verde Archipelago, eastern Atlantic). *Arquipelago* **2000**, *2*, 157–165.

145. Gleibs, S.; Mebs, D.; Werding, B. Studies on the origin and distribution of palytoxin in a Caribbean coral reef. *Toxicon* **1995**, *33*, 1531–1537. [CrossRef]

146. Cortés, J. Biodiversidad marina de Costa Rica: Filo Cnidaria. *Rev. Biol. Trop.* **1997**, *44*, 323–334.

147. Cortés, J.; Murillo, M.M.; Guzmán, H.M.; Acuña, J. Pérdida de zooxantelas y muerte de corales y otros organismos arrecifales en el Caribe y Pacífico de Costa Rica. *Rev. Biol. Trop.* **1984**, *32*, 227–231.

148. Duerden, J.E. Report on the actinians of Puerto Rico. *U. S. Fish. Comm. Bull.* **1902**, *20*, 321–354.

149. Varela, C.; Guitart, B.; Ortiz, M.; Lalana, R. Los zoantideos (Cnidaria, Anthozoa, Zoanthiniaria), de la región occidental de Cuba. *Rev. Investig. Mar.* **2002**, *23*, 179–184.

150. Pax, F. Actiniarien, Zoantharien und Ceriantharien von Curaçao. *Bijdr. Dierk.* **1924**, *23*, 93–121.

151. Goreau, T.F. The ecology of Jamaican coral reefs I. Species composition and zonation. *Ecology* **1959**, *40*, 67–90. [CrossRef]

152. Goreau, T.F. Mass expulsion of zooxanthellae from Jamaican reef communities after Hurricane Flora. *Science* **1964**, *145*, 383–386. [CrossRef]

153. Karlson, R.H. Alternative competitive strategies in a periodically disturbed habitat. *Bull. Mar. Sci.* **1980**, *30*, 894–900.

154. Banaszak, A.T.; Santos, M.G.B.; LaJeunesse, T.C.; Lesser, M.P. The distribution of mycosporine-like amino acids (MAAs) and the phylogenetic identity of symbiotic dinoflagellates in cnidarian hosts from the Mexican Caribbean. *J. Exp. Mar. Biol. Ecol.* **2006**, *337*, 131–146. [CrossRef]

155. LaJeunesse, T. Diversity and community structure of symbiotic dinoflagellates from Caribbean coral reefs. *Mar. Biol.* **2002**, *141*, 387–400. [CrossRef]

156. Fadlallah, Y.H.; Karlson, R.H.; Sebens, K.P. A comparative study of sexual reproduction in three species of Panamanian zoanthids (Coelenterata: Anthozoa). *Bull. Mar. Sci.* **1984**, *35*, 80–89.

157. Sebens, K.P. Intertidal distribution of zoanthids on the Caribbean coast of Panama: Effects of predation and desiccation. *Bull. Mar. Sci.* **1982**, *32*, 316–335.

158. Edmunds, P.J. Patterns in the distribution of juvenile corals and coral reef community structure in St. John, US Virgin Islands. *Mar. Ecol. Prog. Ser.* **2000**, *202*, 113–124. [CrossRef]

159. Haywick, D.W.; Mueller, E.M. Sediment retention in encrusting *Palythoa* spp.—A biological twist to a geological process. *Coral Reefs* **1997**, *16*, 39–46. [CrossRef]

160. Mueller, E.; Haywick, D.W. Sediment assimilation and calcification by the Western Atlantic reef zoanthid, *Palythoa caribaeorum*. *Bull. L'Institut Oceanogr. Monaco* **1995**, *14*, 89–100.

161. Reimer, J.D.; Foord, C.; Irei, Y. Species diversity of shallow water zoanthids (Cnidaria: Anthozoa: Hexacorallia) in Florida. *J. Mar. Biol.* **2012**, *2012*, 856079. [CrossRef]

162. Bastidas, C.; Bone, D. Competitive strategies between *Palythoa caribaeorum* and *Zoanthus sociatus* (Cnidaria: Anthozoa) at a reef flat environment in Venezuela. *Bull. Mar. Sci.* **1996**, *59*, 543–555.

163. Núñez, J.G.; Ariza, L.A.; Jiménez, M. Evaluación de la estructura de las comunidades coralinas en la franja sublitoral de la zona costera sur del Golfo de Cariaco, Venezuela. Parte I: Eje turpialito-quetepe. *Bol. Inst. Oceanogr. Venezuela* **2011**, *50*, 149–159.

164. Ong, C.W.; Reimer, J.D.; Todd, P.A. Morphologically plastic responses to shading in the zoanthids *Zoanthus sansibaricus* and *Palythoa tuberculosa*. *Mar. Biol.* **2013**, *160*, 1053–1064. [CrossRef]

165. Ellis, J.; Solander, D. *The Natural History of Many Curious and Uncommon Zoophytes Collected from Various Parts of the Globe*; Benjamin White and Son: London, UK, 1786.

166. Dana, J.D. Zoophytes. In *United States Exploring Expedition during the Years 1838, 1839, 1840, 1841, 1842*; Dougal, W.H., Stuart, F.D., Wilkes, C., Eds.; C. Sherman: Philadelphia, PA, USA, 1846; pp. 7–113.

167. Milne Edwards, H. *Histoire Naturelle des Coralliaires ou Polypes Proprement Dits*; Librairie encyclopédique de Roret: Paris, France, 1857; Volume 1.

168. Rodríguez-Viera, L.; Rodríguez-Casariego, J.; Pérez-García, J.A.; Olivera, Y.; Perera-Pérez, O. Invertebrados marinos de la zona central del golfo de Ana María, Cuba. *Rev. Investig. Mar.* **2012**, *32*, 30–38.

169. Haddon, A.C.; Shackleton, A.M. Actiniae: I. Zoantheae. In Reports on the Zoological Collections Made in the Torres Straits by Professor, A.C. Haddon, 1888–1889. *Sci. Trans. Roy. Dublin Soc. ser. 2* **1891**, *4*, 658–673, pls. 61–64.

170. Winston, J.E. Diversity and distribution of bryozoans in the Pelican Cays, Belize, Central America. *Atoll Res. Bull.* **2007**, *546*, 1–24. [CrossRef]

171. Schoenberg, D.A.; Trench, R.K. Genetic variation in *Symbiodinium* (=*Gymnodinium*) *microadriaticum* Freudenthal, and specificity in its symbiosis with marine invertebrates. I. Isoenzyme and soluble protein patterns of axenic cultures of *Symbiodinium microadriaticum*. *Proc. Roy. Soc. B* **1980**, *207*, 405–427. [CrossRef]

172. Costa, D.L.; Santos, A.M.; da Silva, A.F.; Padilha, R.M.; Nogueira, V.O.; Wanderlei, E.B.; Bélanger, D.; Gomes, P.B.; Pérez, C.D. Biological impacts of the port complex of Suape on benthic reef communities (Pernambuco–Brazil). *J. Coast. Res.* **2014**, *30*, 362–370. [CrossRef]

173. Cruz, I.C.S.; de Kikuchi, R.K.P.; Longo, L.L.; Creed, J.C. Evidence of a phase shift to *Epizoanthus gabrieli* Carlgreen, 1951 (Order Zoanthidea) and loss of coral cover on reefs in the Southwest Atlantic. *Mar. Ecol.* **2015**, *36*, 318–325. [CrossRef]

174. Cruz, I.C.S.; Loiola, M.; Albuquerque, T.; Reis, R.; José de Anchieta, C.C.; Reimer, J.D.; Mizuyama, M.; Kikuchi, R.K.P.; Creed, J.C. Effect of phase shift from corals to Zoantharia on reef fish assemblages. *PLoS ONE* **2015**, *10*, e0116944. [CrossRef]

175. Kelmo, F.; Attrill, M.J.; Jones, M.B. Effects of the 1997–1998 El Niño on the cnidarian community of a high turbidity coral reef system (northern Bahia, Brazil). *Coral Reefs* **2003**, *22*, 541–550. [CrossRef]

176. Metri, R.; Rocha, R.M. Bancos de algas calcárias, um ecossistema rico a ser preservado. *Natureza & Conservação* **2008**, *8*, 8–17.

177. Soares, M.O.; de Souza, L.P. Osmorregulação no zoantídeo tropical *Protopalythoa variabilis* (Cnidaria: Anthozoa). *Acta Sci. Biol. Sci.* **2013**, *35*, 123–127. [CrossRef]

178. Wilke, D.V.; Jimenez, P.C.; Araújo, R.M.; Pessoa, O.D.L.; Silveira, E.R.; Pessoa, C.; Moraes, M.O.; Lopes, N.P.; Costa-Lotufo, L.V. A new cytotoxic 2-amino-n-alkyl-carboxylic acid mixture obtained from the zoanthid *Protopalythoa variabilis* collected at Paracuru beach, Ceará State, Brazil. *Planta Med.* **2008**, *74*, 1060. [CrossRef]

179. Wilke, D.V.; Jimenez, P.C.; Pessoa, C.; de Moraes, M.O.; Araújo, R.M.; da Silva, W.M.B.; Silveira, E.R.; Pessoa, O.D.L.; Braz-Filho, R.; Lopes, N.P. Cytotoxic lipidic α-amino acids from the zoanthid *Protopalythoa variabilis* from the northeastern coast of Brazil. *J. Braz. Chem. Soc.* **2009**, *20*, 1455–1459. [CrossRef]

180. Diop, M.; Leug-Tack, D.; Braekman, J.C.; Kornprobst, J.M. Sterol composition of four Zoathidae members of the genus *Palythoa* from the Cape Verde Peninsula. *Biochem. Syst. Ecol.* **1986**, *14*, 151–154. [CrossRef]

181. Karlson, R.H. Disturbance and monopolization of a spatial resource by *Zoanthus sociatus* (Coelenterata, Anthozoa). *Bull. Mar. Sci.* **1983**, *33*, 118–131.

182. Koehl, M.A. Water flow and the morphology of zoanthid colonies. In Proceedings of the 3rd International Coral Reef Symposium, Miami, FL, USA, May 1977; Taylor, D.L., Ed.; International Coral Reef Society: Miami, FL, USA, 1977; pp. 437–444.

183. Lamarck, J.B.P. *Système des Animaux sans Vertèbres*; Self-Published: Paris, France, 1801.

184. Villar, R.M.; Gil-Longo, J.; Daranas, A.H.; Souto, M.L.; Fernandez, J.J.; Peixinho, S.; Barral, M.A.; Santafe, G.; Rodríguez, J.; Jiménez, C. Evaluation of the effects of several zoanthamine-type alkaloids on the aggregation of human platelets. *Bioorg. Med. Chem.* **2003**, *11*, 2301–2306. [CrossRef]

185. Ellis, J. An account of the Actinia Sociata, or clustered animal-flower, lately found on the sea-coasts of the new-ceded islands: In a letter from John Ellis, Esquire, F.R.S. to the Right Honourable the Earl of Hillsborough, F.R.S. *Phil. Trans.* **1768**, *57*, 428–437.

186. McMurrich, J.P. Notes on some Actinians from the Bahama Islands, collected by the late Dr. J.I. Northrop. *Ann. NY Acad. Sci.* **1896**, *IX*, 181–194. [CrossRef]

187. McMurrich, J.P. Report on the Actiniaria collected by the Bahama Expedition of the State University of Iowa, 1893. *Bull. Lab. Nat. Hist. State Univ. Iowa* **1898**, *4*, 225–249, pls. 1–3.

188. Laborel, J. Les peuplements de Madréporaires des côtes tropicales du Brésil. *Ann. Univ. Abidjan* **1970**, *2*, 1–260.

189. De O. Pires, D.; Migotto, A.E.; Marques, A.C. Cnidários bentônicos do Arquipélago de Fernando de Noronha, Brasil. *Boletim Museu Nac. Rio Jan. NS Zool.* **1992**, *354*, 1–21.

190. Rohlfs de Macedo, C.M.R.; Belém, M.J.C. The genus *Zoanthus* in Brazil. 1. Characterization and anatomical revision of *Zoanthus sociatus* (Cnidaria, Zoanthinaria, Zoanthidae). *Iheringia* **1994**, *77*, 135–144.

191. Sarmento, F.; Correia, M.D. Description of ecological and external morphological parameters of Porifera at the Ponta Verde coral reef, Alagoas, *Brazil*. *Rev. Bras. Zoociênc.* **2002**, *4*, 215–226.

192. Le Sueur, C.A. Observations on several species of the genus *Actinia*: Illustrated by figures. *J. Acad. Nat. Sci. Phil.* **1817**, *1*, 149–189.

193. Karlson, R.H. Fission and the dynamics of genets and ramets in clonal cnidarian populations. In *Coelenterate Biology: Recent Research on Cnidaria and Ctenophora, Developments in Hydrobiology*; Williams, R.B., Cornelius, P.F.S., Hughes, R.G., Robson, E.A., Eds.; Springer: Dordrecht, Germany, 1991; pp. 235–240.

194. Carlgren, O. *Ostafrikanische Actinien, gesammelt von Herrn Dr. F. Stuhlmann 1898 und 1899*, 1st ed.; Mitteilungen aus dem Naturhistorischen Museum in Hamburg: Hamburg, Germany, 1900; pp. 21–144.

195. Kamezaki, M.; Higa, M.; Hirose, M.; Suda, S.; Reimer, J.D. Different zooxanthellae types in populations of the zoanthid *Zoanthus sansibaricus* along depth gradients in Okinawa, Japan. *Mar. Biodivers.* **2013**, *43*, 61–70. [CrossRef]

196. Karlson, R.H. Size-dependent growth in two zoanthid species: A contrast in clonal strategies. *Ecology* **1988**, *69*, 1219–1232. [CrossRef]

197. Gray, J.E. *Spicilegia Zoologica: Original Figures and Short Systematic Descriptions of New and Unfigured Animals. Part 1*; Treüttel, Würtz and Co.: London, UK, 1828.

198. Brown, J.; Downes, K.; Mrowicki, R.J.; Nolan, E.L.; Richardson, A.J.; Swinnen, F.; Wirtz, P. New records of marine invertebrates from Ascension Island (Central Atlantic). *Arquipélago* **2016**, *33*, 71–79.

199. Larson, K.S.; Larson, R.J. On the ecology of *Isaurus duchassaingi* (Andres) (Cnidaria: Zoanthidea) from South Water Cay, Belize. *Smithson. Contrib. Mar. Sci.* **1982**, *12*, 475–488.

200. Grohman, P.A.; Peixinho, S. *Isaurus tuberculatus* (Cnidaria, Anthozoa, Zoanthidea), nova ocorrência para o Atlântico sudoeste Tropical. *Nerítica* **1995**, *9*, 19–22.

201. Rabelo, E.F.; Matthews-Cascon, H. Influence of light on the feeding behaviour of *Isaurus tuberculatus* Gray, 1828 (Cnidaria: Zoanthidea) under laboratory conditions. *Arquivos Ciênc. Mar.* **2007**, *40*, 55–58.

202. Riera, R.; Becerro, M.A.; Stuart-Smith, R.D.; Delgado, J.D.; Edgar, G.J. Out of sight, out of mind: Threats to the marine biodiversity of the Canary Islands (NE Atlantic Ocean). *Mar. Pollut. Bull.* **2014**, *86*, 9–18. [CrossRef] [PubMed]

203. Muirhead, A.; Ryland, J.S. A review of the genus *Isaurus* Gray, 1828 (Zoanthidea), including new records from Fiji. *J. Nat. Hist.* **1985**, *19*, 323–335. [CrossRef]

204. Reimer, J.D.; Ono, S.; Tsukahara, J.; Iwase, F. Molecular characterization of the zoanthid genus *Isaurus* (Anthozoa: Hexacorallia) and associated zooxanthellae (*Symbiodinium* spp.) from Japan. *Mar. Biol.* **2008**, *153*, 351–363. [CrossRef]

205. Cardinale, B.J.; Duffy, J.E.; Gonzalez, A.; Hooper, D.U.; Perrings, C.; Venail, P.; Narwani, A.; Mace, G.M.; Tilman, D.; Wardle, D.; et al. Biodiversity loss and its impact on humanity. *Nature* **2012**, *489*, 326. [CrossRef]

206. Hughes, T.P. Catastrophes, phase shifts, and large-scale degradation of a Caribbean coral reef. *Science* **1994**, *265*, 1547–1551. [CrossRef] [PubMed]

207. Burke, L.; Maidens, J. *Reefs at Risk in the Caribbean*; World Resources Institute (WRI): Washington, DC, USA, 2004.

208. Mora, C. A clear human footprint in the coral reefs of the Caribbean. *Proc. Roy. Soc. B Biol. Sci.* **2008**, *275*, 767–773. [CrossRef] [PubMed]

209. Miloslavich, P.; Díaz, J.M.; Klein, E.; Alvarado, J.J.; Díaz, C.; Gobin, J.; Escobar-Briones, E.; Cruz-Motta, J.J.; Weil, E.; Cortes, J.; et al. Marine biodiversity in the Caribbean: Regional estimates and distribution patterns. *PLoS ONE* **2010**, *5*, e11916. [CrossRef]

210. Jackson, J.B.C.; Donovan, M.K.; Cramer, K.L.; Lam, V.V. *Status and Trends of Caribbean Coral Reefs: 1970–2012*; Global Coral Reef Monitoring Network, IUCN: Gland, Switzerland, 2014.

211. Oliver, L.M.; Fisher, W.S.; Fore, L.; Smith, A.; Bradley, P. Assessing land use, sedimentation, and water quality stressors as predictors of coral reef condition in St. Thomas, U.S. Virgin Islands. *Environ. Monitor. Assess.* **2018**, *190*, 1–25. [CrossRef]

212. Hoeksema, B.W.; van der Land, J.; van der Meij, S.E.T.; van Ofwegen, L.P.; Reijnen, B.T.; van Soest, R.W.M.; de Voogd, N.J. Unforeseen importance of historical collections as baselines to determine biotic change of coral reefs: The Saba Bank case. *Mar. Ecol.* **2011**, *32*, 135–141. [CrossRef]

213. Rocha, L.A.; Aleixo, A.; Allen, G.; Almeda, F.; Baldwin, C.C.; Barclay, M.V.L.; Bates, J.M.; Bauer, A.M.; Benzoni, F.; Berns, C.M.; et al. Specimen collection: An essential tool. *Science* **2014**, *344*, 814–815. [CrossRef]

214. Swain, T.D. Evolutionary transitions in symbioses: Dramatic reductions in bathymetric and geographic ranges of Zoanthidea coincide with loss of symbioses with invertebrates. *Mol. Ecol.* **2010**, *19*, 2587–2598. [CrossRef]

215. Floeter, S.R.; Rocha, L.A.; Robertson, D.R.; Joyeux, J.C.; Smith-Vaniz, W.F.; Wirtz, P.; Edwards, A.J.; Barreiros, J.P.; Ferreira, C.E.L.; Gasparini, J.L.; et al. Atlantic reef fish biogeography and evolution. *J. Biogeogr.* **2008**, *35*, 22–47. [CrossRef]

216. Veron, J.; Stafford-Smith, M.; DeVantier, L.; Turak, E. Overview of distribution patterns of zooxanthellate Scleractinia. *Front. Mar. Sci.* **2015**, *1*, 1–19. [CrossRef]

217. Sinniger, F.; Ocaña, O.V.; Baco, A.R. Diversity of zoanthids (Anthozoa: Hexacorallia) on Hawaiian seamounts: Description of the Hawaiian gold coral and additional zoanthids. *PLoS ONE* **2013**, *8*, e52607. [CrossRef] [PubMed]

218. Reimer, J.D.; Kise, H.; Santos, M.E.; Lindsay, D.J.; Pyle, R.L.; Copus, J.M.; Bowen, B.W.; Nonaka, M.; Higashiji, T.; Benayahu, Y. Exploring the biodiversity of understudied benthic taxa at mesophotic and deeper depths: Examples from the order Zoantharia (Anthozoa: Hexacorallia). *Front. Mar. Sci.* **2019**, *6*, 305. [CrossRef]

219. Reimer, J.D.; Poliseno, A.; Hoeksema, B.W. Shallow-water zoantharians (Cnidaria, Hexacorallia) from the Central Indo-Pacific. *ZooKeys* **2014**, *444*, 1–57. [CrossRef] [PubMed]

220. Reimer, J.D.; Wee, H.B.; Ang, P.; Hoeksema, B.W. Zoantharia (Cnidaria: Anthozoa: Hexacorallia) of the South China Sea and Gulf of Thailand: A species list based on past reports and new photographic records. *Raffles Bull. Zool.* **2015**, *63*, 334–356.

 *diversity*

Communication

Widespread Occurrence of a Rarely Known Association between the Hydrocorals *Stylaster roseus* and *Millepora alcicornis* at Bonaire, Southern Caribbean

Simone Montano [1,2,*], James D. Reimer [3,4], Viatcheslav N. Ivanenko [5], Jaaziel E. García-Hernández [6], Godfried W.N.M. van Moorsel [7,8], Paolo Galli [1,2] and Bert W. Hoeksema [9,10]

[1] Department of Earth and Environmental Sciences (DISAT), University of Milan – Bicocca, Piazza della Scienza, 20126 Milan, Italy; paolo.galli@unimib.it
[2] MaRHE Center (Marine Research and High Education Center), Magoodhoo Island, 12030 Faafu Atoll, Maldives
[3] Molecular Invertebrate Systematics and Ecology Laboratory, Graduate School of Engineering and Science, University of the Ryukyus, Okinawa 903-0213, Japan; jreimer@sci.u-ryukyu.ac.jp
[4] Tropical Biosphere Research Center, University of the Ryukyus, Okinawa 903-0213, Japan
[5] Department of Invertebrate Zoology, Biological Faculty, Lomonosov Moscow State University, 119992 Moscow, Russia; ivanenko.slava@gmail.com
[6] Marine Genomic Biodiversity Laboratory, University of Puerto Rico - Mayagüez, La Parguera, PR 00667, USA; jaaziel.garcia@upr.edu
[7] Ecosub, Berkenlaantje 2, 3956 DM Leersum, The Netherlands; vanmoorsel@ecosub.nl
[8] ANEMOON Foundation, P.O. Box 29, 2120 AA Bennekom, The Netherlands
[9] Taxonomy and Systematics Group, Naturalis Biodiversity Center, P.O. Box 9517, 2300 RA Leiden, The Netherlands; bert.hoeksema@naturalis.nl
[10] Groningen Institute for Evolutionary Life Sciences, University of Groningen, P.O. Box 11103, 9700 CC Groningen, The Netherlands
* Correspondence: simone.montano@unimib.it; Tel.: +39-02-6448-2050

Received: 14 May 2020; Accepted: 29 May 2020; Published: 30 May 2020

Abstract: Among symbiotic associations, cases of pseudo-auto-epizoism, in which a species uses a resembling but not directly related species as substrate, are poorly documented in coral reef ecosystems. In the present study, we assessed the distribution of an association between the hydrocorals *Stylaster roseus* and *Millepora alcicornis* on about 50% of coral reef sites studied in Bonaire, southern Caribbean. Although previously thought to be uncommon, associations between the lace coral *S. roseus* and the fire coral *M. alcicornis* were observed at both the windward and leeward sides of Bonaire, mainly between 15 and 25 m depth, reaching a maximum occupation of 47 *S. roseus* colonies on a single *M. alcicornis* colony. Both species' tissues did not show any signs of injuries, while an in-depth inspection of the contact points of their skeletons revealed that both partners can partially overgrow each other. How it is possible that *S. roseus* is able to settle on the stinging tissue of *Millepora* as well as how, by contrast, the latter may facilitate the lace coral by offering a certain degree of protection are questions that deserve further investigations.

Keywords: Caribbean Netherlands; fire corals; Hydrozoa; pseudo-auto-epizoism; stony corals; substrate; symbiosis

1. Introduction

As one of the most species-rich marine ecosystems, coral reefs are renowned for their plethora of different interspecific associations [1]. Reef-building corals serve as home for diverse assemblages of macro- and micro-organisms from all kingdoms of life [2], which rely on them for food, shelter and substrate [3–5].

To date, reef-building corals belonging to the order Scleractinia account for the majority of host species studied [3], although other hosts of other benthic groups exist, such us anthozoans, bryozoans, sponges, ascidians and hydrocorals.

Regarding hydrocorals (Class Hydrozoa), the genus *Millepora* Linnaeus, 1758 (Suborder Capitata) occurs as a circumtropical component of shallow-water coral reefs [6–11]. Milleporids are found in depths of less than 1 m to about 40 m and provide substratum for sedentary organisms and food or shelter for mobile ones [12–16]. The three-dimensional structural complexity of the colonies generated by in particular branching *Millepora* species harbors a great diversity of organisms that includes crustaceans, worms, fishes and other organisms that live in close association [12–14].

Similar to *Millepora*, hydrocorals belonging to the family Stylasteridae (Suborder Filifera), commonly known as "lace corals", are colonial hydroids characterized by a calcium carbonate skeleton, with 90% of the species occurring at depths over 50 m [17]. Their colonies are generally erect and branching with only the Pacific genus *Stylantheca* Fisher, 1931 having an encrusting morphology [18]. Most lace corals, including those of the genus *Stylaster* Gray, 1831, are known to form strict relationships with other invertebrates [18,19]. Stylasterids are host of a number of commensals such as polychaetes [20–22], nemerteans, pycnogonids, cirripids, barnacles, and bryozoans [23,24]. In addition to these, six species of gall living siphonostomatoid copepods of the family Asterocheridae [25,26] and tiny ovulid gastropods of the genus *Pedicularia* Swainson, 1840 are known to be obligate symbionts on various stylasterid species [27,28].

Among interspecific relationships involving hydrozoans, a minority includes a hydrozoan-hydrozoan association that has been classified as auto-epizoism, even though the two partners do not belong to the same species [29,30]. Indeed, several hydroids are known to live epizootically on other hydroids, usually using the other as a solid substratum, e.g., members of the genera *Hebella* Allman, 1888 and *Anthohebella* Boero, Bouillon & Kubota 1997 can be observed on the perisarc of other hydroid species [31,32].

Currently no information is present about a possible association between *Stylaster* and *Millepora*, although two earlier observations have been reported from the southern and eastern Caribbean [33,34]. Here we investigated the distribution and abundance of the association between *Stylaster roseus* (Pallas, 1766) and *Millepora alcicornis* Linnaeus, 1758 found during a biodiversity expedition conducted at Bonaire, in the Dutch Caribbean. In addition, an in-depth morphological analysis of the skeleton interactions between both partners is provided.

2. Materials and Methods

The study was conducted in the waters of Bonaire from October to November 2019. We explored 34 localities, chosen randomly among accessible sites (Figure 1; Table S1). The presence of the *Stylaster-Millepora* association was recorded by applying the roving diving technique with scuba, in which a 1-h dive served as the sampling unit, by starting at the maximum depth at each dive locality (15–35 m) and moving to shallower water from there [35]. Even though the information collected from this timed dive method does not result in quantitative data per site, it is particularly useful when the goal of the study is to compare biodiversity among site via finding as many species as possible, or when looking for small and/or cryptic species and symbiotic associations [36]. Moreover, to preliminarily assess the abundance and spatial distribution of this association, the total number of *Stylaster-Millepora* associations and their depth ranges were recorded for each site. In addition, the total numbers of colonies of *S. roseus* found to grow on each *M. alcicornis* colony were noted. For documentation purposes underwater photographs of the *Stylaster-Millepora* association were taken using a Canon GX7 Mark II camera in a Fantasea GX7 II underwater housing and a Nikon D7100 in a Hugyfot housing.

A branch fragment of about 10 cm in length of *M. alcicornis* colonized by *S. roseus* was collected for further analyses. Microphotographs (×32) of stylasterids growing on the coral skeletons of *M. alcicornis* were taken using a Leica EZ4 D stereo microscope equipped with a Canon GX7 Mark II camera. Several parts of the fragment were observed using the Zeiss Gemini SEM500 scanning electron microscope operating at beam energies of 5 kV (ZEISS, Oberkochen, Germany) in order to characterize the skeletal interface/interactions of the association.

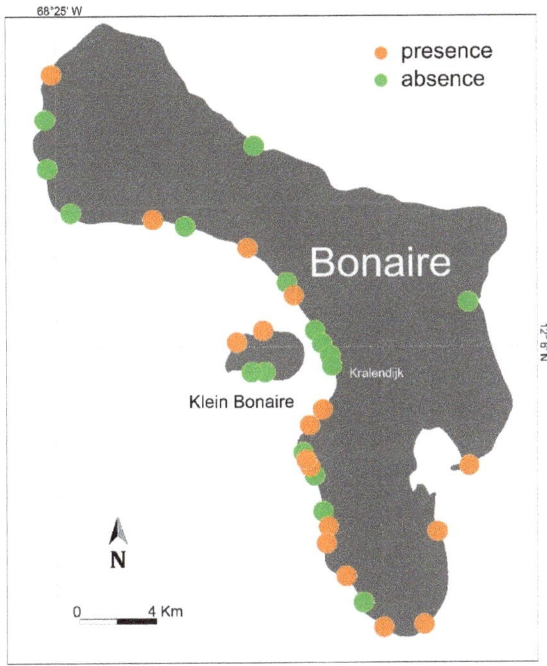

Figure 1. Map of Bonaire island showing locations where the *Stylaster-Millepora* association was found among all sites surveyed.

3. Results and Discussion

Bonaire is known for its rich coral reef ecosystem, considered as one of the healthiest and most resilient in the Caribbean [37], and therefore it serves as a major tourist destination for scuba divers and snorkellers. In spite of being one of the most popular diving spots in the Caribbean [38,39], this is the first time that the *Stylaster-Millepora* association has been reported. Moreover, to the best of our knowledge, this is the first report that preliminarily assesses abundance, depth distribution and skeleton interactions involving both partners.

Our biodiversity survey revealed that the *Stylaster-Millepora* association was found at 17 out of 34 (50%) of the sites explored (Figure 1; Table S1). Interestingly, this apparently uncommon association seems to be relatively abundant and widespread on Bonaire's coral reefs. In fact, it was observed at both the windward and leeward sides of Bonaire, despite the reefs along the island' windward shores being generally less developed compared to those of the leeward coast [40].

In particular, *S. roseus* and *M. alcicornis* were observed to form strict relationships (Figure 2a) in depths ranging from 13 to 32 m. In total we counted 55 colonies of *M. alcicornis* hosting *S. roseus* with the highest number of records ($n = 37$) between 15–25 m depths. Again, the density of *S. roseus* colonies on *M. alcicornis* colonies ranged from one to at least 47 for the largest fire coral colonies (Figure 2b; Table S1). The observed pattern is not surprising since a high diversity of associated invertebrates has been reported from the complex structure of *M. alcicornis*, showing high numbers of individuals, which appear to be directly related to the volume of the host colony [14]. Moreover, additional observations made using a stereo microscope revealed the presence of newly settled single-cyclosystem colonies (Figure 2c), which were not observable with the naked eye. This suggests that the number of *S. roseus* colonies present on each single *M. alcicornis* coral is probably higher than observed. Because *Millepora* and *Stylaster* represent different families of hydrocorals, Milleporidae and Stylasteridae, that only resemble each other by forming a calcareous skeleton, while both belong to different suborders, their relation is not considered an example of auto-epizoism but the first case of what can be considered pseudo-auto-epizoism.

Figure 2. *Stylaster-Millepora* association. (**a**) Overview of field appearance; (**b**) high number of *S. roseus* colonies growing on a *M. alcicornis* colony; (**c**) a single *S. roseus* recruit on *M. alcicornis*; (**d**) exposure of *S. roseus* when overgrowing a fire coral colony; (**e,f**) peculiar pattern of *S. roseus* and *M. alcicornis* growth and close-up of both colonies with extended polyps.

Exploring the position of the association we noted that most *S. roseus* colonies were clearly located in the upper parts or in windows between the anastomose branches of *M. alcicornis*, while they were always abnormally exposed (Figure 2d). This is remarkable because, typically, *S. roseus* is a very cryptic species, inhabiting shaded crevices of the reef or the dead underside of foliaceous corals [33]. Occasionally they are found underneath overhangs or large rocks in shallow water (<3 m depth) where the colonies are more exposed to light and wave action (Figure S1). However, in these particular cases, when associated with *M. alcicornis*, the lace corals are completely exposed with colonies protruding upwards or toward the maximum exposition of water motion.

In addition, another interesting pattern was observed. In the portions of the colonies of *M. alcicornis* where *S. roseus* are located both species appear to grow in synchrony by creating a specular shape of the main three-dimensional structure of the colonies (Figure 2e). In this case, if it is *S. roseus* that limits the growth of *M. alcicornis* or, in contrast, if *S. roseus* simply fills the gaps between the anastomosing *Millepora* branches, both seem to be valid hypotheses. In this context, we highlight how, in the majority of the cases, the polyps of *S. roseus* and *M. alcicornis* were frequently observed extended at the same time (Figure 2f). However, in all our observations no evidence of injuries to either species were detected. In fact, in this scenario, both partners do not seem to trigger the stinging cells of polyps of their partner species despite potentially being in contact. It may be that the well-known stinging properties of *Millepora* fire corals may facilitate this association by conferring a certain degree of protection. This is currently unknown and worthy of further investigations.

The patterns observed at the edge of both tissues also deserve attention. In this case, no apparent inflammation status or stressed condition related to adjacent growth of the partner was found (Figure 3a). In contrast, on a macroscopic scale, the tissues of both partners seem to attach so tightly that the boundaries between them almost appear to have disappeared (Figure 3b).

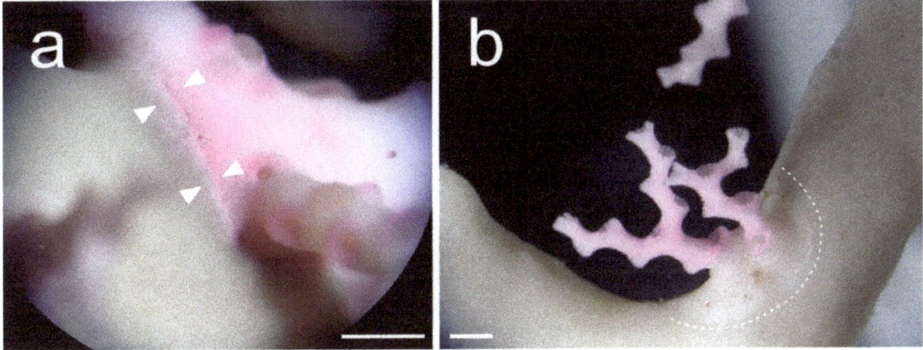

Figure 3. (**a**) Apparently healthy tissue of both partners when intimately close, and (**b**) the undefined borders/edges between the two hydrocorals' tissues. (Scale bars ~ 2 mm).

An in-depth inspection of the skeletons showed that the skeleton tissue of *M. alcicornis* can overgrow that of *S. roseus* and may limit its growth (Figure 4a,b). In contrast, in some other observations it was evident that the skeleton tissue of *S. roseus* also easily overgrows that of *M. alcicornis*, extending the colony above the *M. alcicornis* tissue and producing an enlarged basal disc from which new cyclosystems arise (Figure 4c,d). Although both mechanisms can be deducted by the observation of the growth patterns of the skeletons, we found also a portion of the interface between the two colonies that appeared literally fused, on which the edges of both species were almost undefined (Figure 4e,f).

In colonies of *M. alcicornis*, the stinging cells in the epidermis can form a barrier against the larval settlement [33], but this does not seem to prevent settlement of new *S. roseus* recruits. The early development stages of the new coenosteum after planulae settlement are known in only a few stylasterid species. Ostarello [41] studied the natural history of *Stylaster* sp. and observed that after release, the planula generally crawled for a short time around the parental colony before settling close to it. How *S. roseus* larvae can settle on *M. alcicornis* in the present case is not understood.

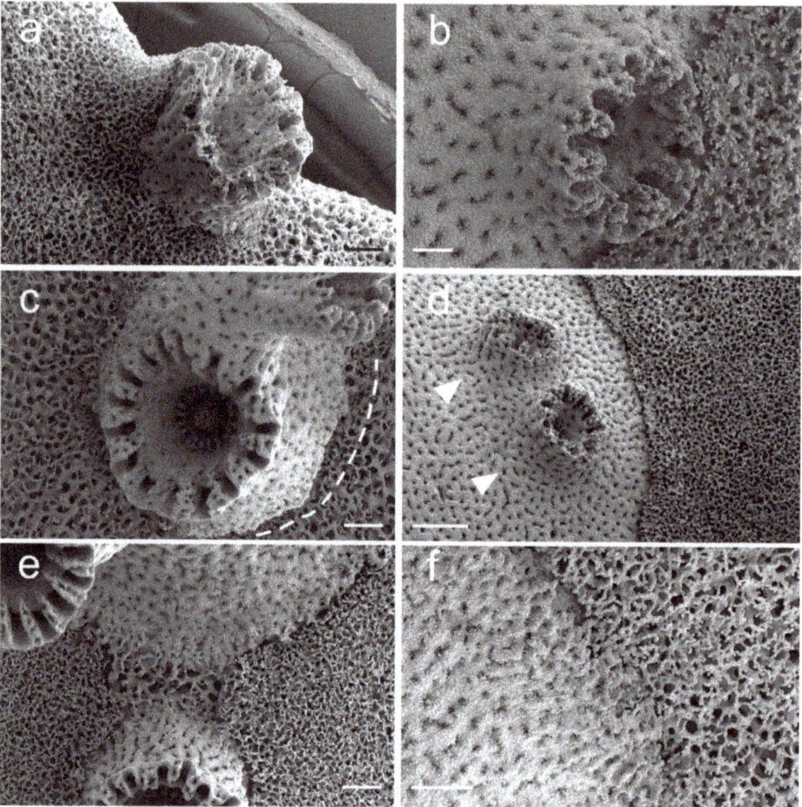

Figure 4. SEM analyses revealed (**a,b**) the capacity of *Millepora alcicornis* to overgrowth *Stylaster roseus*; (**c,d**) the capacity of *S. roseus* to overgrowth on *M. alcicornis*; (**e,f**) the apparently intimately connection of both skeletons in some parts of the associations. (Scale bars: a–c ~100 µm; d ~200 µm; e–f ~100 µm).

Furthermore, it would be intriguing to investigate the advantages of *S. roseus* in colonizing *M. alcicornis*. In fact, despite the fact that this fire coral species has many different growth morphologies [42,43] (Figure S2a,b), with branches often becoming anastomose (Figure S2c) and crucial in providing many different places where *S. roseus* may settle, the higher growth rate of *M. alcicornis* [44] may also results in a total overgrowth of *S. roseus* colonies (Figure S2d).

By contrast, the "sheet-tree" morphology of *M. alcicornis* (sensu [44]) appears to have a number of beneficial consequences as the role in competitive interactions, zooplanktivory, and asexual reproduction [45] that may be potentially vital for the lace coral. Thus, we cannot exclude that *S. roseus* may exploit *M. alcicornis* to increase its capture of zooplankton, taking advantages of the greater asexual reproduction of *M. alcicornis* in order to support its natural turnover, as well as to increase the spatial diffusion of the species and to reduce the conspecific competition.

In conclusion, it will be necessary to understand how this association affects both partners and how it is of benefit to both of them. Although elucidating the nature of the diverse types of symbiotic interactions is not always easy, it has already been demonstrated that symbionts may play an active role in protecting their hosts from various stresses [46–48]. Further investigations on the nature of this *S. roseus* and *M. alcicornis* association will undoubtedly provide more insights into the nature of symbioses on coral reefs.

Moreover, since we cannot rule out the possibility that our results may not be representative of large-scale patterns valid for the whole Bonaire reef, we hope that our preliminary data will promote future in-depth ecological investigations focusing on this association.

Supplementary Materials: The following are available online at http://www.mdpi.com/1424-2818/12/6/218/s1, Figure S1. *Stylaster rosesus* in its natural habitat. Figure S2. Different growth morphologies of *M. alcicornis*. Table S1. List of locations and number of associations recorded for each site sampled.

Author Contributions: Conceptualization, S.M. and B.W.H.; methodology, S.M.; investigation, S.M., J.D.R., V.N.I., J.E.G.-H.; resources, B.W.H., P.G., G.W.N.M.v.M.; data curation, S.M.; writing—original draft preparation, S.M., B.W.H., J.D.R.; All authors have read and agreed to the published version of the manuscript.

Funding: The fieldwork in Bonaire was supported by a grant from the WWF-Netherlands Biodiversity Fund to BWH.

Acknowledgments: S.M., J.D.R., V.N.I., and J.E.G.-H. are grateful to Naturalis Biodiversity Center for supporting fieldwork in Bonaire (2019). We are grateful to STINAPA and DCNA at Bonaire for assistance in the submission of the research proposal and the research permit. S.M. would like to express his sincere gratitude to Morena Nava for the precious support provided.

Conflicts of Interest: The authors declare no conflict of interest.

References

1. Gates, R.D.; Ainsworth, T.D. The nature and taxonomic composition of coral symbiomes as drivers of performance limit in scleractinian corals. *J. Exp. Mar. Biol. Ecol.* **2011**, *408*, 94–101. [CrossRef]

2. Rohwer, F.; Breitbart, M.; Jara, J.; Azam, F.; Knowlton, N. Diversity of bacteria associated with the Caribbean coral *Montastraea franksi*. *Coral Reefs* **2001**, *20*, 85–91.

3. Stella, J.S.; Pratchett, M.S.; Hutchings, P.A.; Jones, G.P. Coral-associated invertebrates: Diversity, ecological importance and vulnerability to disturbance. *Oceanogr. Mar. Biol. Annu. Rev.* **2011**, *49*, 43–116.

4. Hoeksema, B.W.; van der Meij, S.E.T.; Fransen, C.H.J.M. The mushroom coral as a habitat. *J. Mar. Biol. Assoc. U. K.* **2012**, *92*, 647–663. [CrossRef]

5. Hoeksema, B.W. The hidden biodiversity of tropical coral reefs. *Biodiversity* **2017**, *18*, 8–12. [CrossRef]

6. De Souza, J.N.; Nunes, F.L.D.; Zilberberg, C.; Sanchez, J.A.; Migotto, A.E.; Hoeksema, B.W.; Serrano, X.M.; Baker, A.C.; Lindner, A. Contrasting patterns of connectivity among endemic and widespread fire coral species (*Millepora* spp.) in the tropical Southwestern Atlantic. *Coral Reefs* **2017**, *36*, 701–716. [CrossRef]

7. Arrigoni, R.; Maggioni, D.; Montano, S.; Hoeksema, B.W.; Seveso, D.; Shlesinger, T.; Terraneo, T.I.; Tietbohl, M.D.; Berumen, M.I. An integrated morpho-molecular approach to delineate species boundaries of *Millepora* from the Red Sea. *Coral Reefs* **2018**, *37*, 967–984. [CrossRef]

8. Takama, O.; Fernandez-Silva, I.; López, C.; Reimer, J.D. Molecular phylogeny demonstrates the need for taxonomic reconsideration of species diversity of the hydrocoral genus *Millepora* (Cnidaria: Hydrozoa) in the Pacific. *Zool. Sci.* **2018**, *35*, 123–133. [CrossRef]

9. Boissin, E.; Pogoreutz, C.; Pey, A.; Gravier-Bonnet, N.; Planes, S. *Millepora platyphylla* (Cnidaria, Hydrozoa) range extended back to the Eastern Pacific, thanks to a new record from Clipperton Atoll. *Zootaxa* **2019**, *4668*, 599–600. [CrossRef]

10. Rodríguez, L.; López, C.; Casado-Amezua, P.; Ruiz-Ramos, D.V.; Martínez, B.; Banaszak, A.; Tuya, F.; García-Fernández, A.; Hernández, A. Genetic relationships of the hydrocoral *Millepora alcicornis* and its symbionts within and between locations across the Atlantic. *Coral Reefs* **2019**, *38*, 255–268. [CrossRef]

11. Boissin, E.; Leung, J.K.L.; Denis, V.; Bourmaud, C.A.F.; Gravier-Bonnet, N. Morpho-molecular delineation of structurally important reef species, the fire corals, *Millepora* spp., at Réunion Island, Southwestern Indian Ocean. *Hydrobiologia* **2020**, *847*, 1237–1255. [CrossRef]

12. Lewis, J.B. Biology and ecology of the hydrocoral *Millepora* on coral reefs. *Adv. Mar. Biol.* **2006**, *50*, 1–55. [PubMed]

13. Garcia, T.M.; Matthews-Cascon, H.; Franklin-Junior, W. Macrofauna associated with branching fire coral *Millepora alcicornis* (Cnidaria: Hydrozoa). *Thalassas* **2008**, *24*, 11–19.

14. Garcia, T.M.; Matthews-Cascon, H.; Franklin-Junior, W. *Millepora alcicornis* (Cnidaria: Hydrozoa) as substrate for benthic fauna. *Braz. J. Oceanog.* **2009**, *57*, 153–155. [CrossRef]

15. Hoeksema, B.W.; Ten Hove, H.A. The invasive sun coral *Tubastraea coccinea* hosting a native Christmas tree worm at Curaçao, Dutch Caribbean. *Mar. Biodivers.* **2017**, *47*, 59–65. [CrossRef]

16. Hoeksema, B.W.; García-Hernández, J.E. Host-related morphological variation of dwellings inhabited by the crab *Domecia acanthophora* in the corals *Acropora palmata* and *Millepora complanata* (Southern Caribbean). *Diversity* **2020**, *12*, 143. [CrossRef]

17. Cairns, S.D. Deep-water corals: An overview with special reference to diversity and distribution of deep-water scleractinian corals. *Bull. Mar. Sci.* **2007**, *81*, 311–322.

18. Cairns, S.D. Global diversity of the Stylasteridae (Cnidaria: Hydrozoa: Athecatae). *PLoS ONE* **2011**, *6*, e21670. [CrossRef]

19. Pica, D.; Cairns, S.D.; Puce, S.; Newman, W.A. Southern hemisphere deep-water stylasterid corals including a new species, *Errina labrosa* sp. n. (Cnidaria, Hydrozoa, Stylasteridae), with notes on some symbiotic scalpellids (Cirripedia, Thoracica, Scalpellidae). *ZooKeys* **2015**, *472*, 1–25. [CrossRef]

20. Light, W.J. *Polydora alloporis*, new species, a commensal spionid (Annelida, Polychaeta) from a hydrocoral off Central California. *Proc. Cal. Acad. Sci.* **1970**, *37*, 459–472.

21. Pettibone, M.H. Polynoid commensals with gorgonian and stylasterid corals, with a new genus, new combinations, and new species (Polychaeta: Polynoidae: Polynoinae). *Proc. Biol. Soc. Wash.* **1991**, *104*, 688–713.

22. Martin, D.; Britayev, T.A. Symbiotic polychaetes revisited: An update of the known species and relationships (1998–2017). *Oceanogr. Mar. Biol. Ann. Rev.* **2018**, *56*, 371–448.

23. Zibrowius, H. Associations of Hydrocorallia Stylasterina with gall-inhabiting Copepoda Siphonostomatoidea from the South-West Pacific. Part 1. On the stylasterine hosts, including two new species, *Stylaster papuensis* and *Crypthelia cryptotrema*. *Bijdr Dierk* **1981**, *51*, 268–286.

24. Zibrowius, H.; Cairns, S.D. Revision of the Northeast Atlantic and Mediterranean Stylasteridae (Cnidaria: Hydrozoa). *Mém Mus Nat Hist Nat. Ser. A* **1992**, *153*, 1–136.

25. Stock, J.H. Associations of Hydrocorallia Stylasterina with gall-inhabiting Copepoda Siphonostomatoidea from the South-West Pacific. Part II. On six species belonging to four new genera of the copepod family Asterocheridae. *Bijdr. Dierk.* **1981**, *51*, 287–312.

26. Stock, J.H. On the presence of gall-inducing Copepoda on stylasterine corals.-Proceedings of the First International Conference on Copepoda. Studies on Copepoda II. *Crustaceana* **1984**, *7*, 377–380.

27. Goud, J.; Hoeksema, B.W. Pedicularia vanderlandi spec. nov., a symbiotic snail (Caenogastropoda; Ovulidae) on the hydrocoral *Distichopora vervoorti* Cairns and Hoeksema, 1998 (Hydrozoa; Stylasteridae), from Bali, Indonesia. *Zool. Verh.* **2001**, *334*, 77–97.

28. Wisshak, M.; López Correa, M.; Zibrowius, H.; Jakobsen, J.; Freiwald, A. Skeletal reorganisation affects geochemical signals, exemplified in the stylasterid hydrocoral *Errina dabneyi* (Azores Archipelago). *Mar. Ecol. Prog. Ser.* **2009**, *397*, 197–208. [CrossRef]

29. Millard, N.A.H. Auto-epizoism in South African hydroids. In: Recent trends in research in Coelenterate biology. Proceedings of the Second International Symposium on Cnidaria. *Publ. Seto Mar. Biol. Lab.* **1973**, *20*, 23–34. [CrossRef]

30. Galea, H.R.; Leclère, L. On some morphologically aberrant, auto-epizootic forms of *Plumularia setacea* (Linnaeus, 1758) (Cnidaria: Hydrozoa) from southern Chile. *Zootaxa* **2007**, *1484*, 39–49. [CrossRef]

31. Boero, F.; Bouillon, J.; Kubota, S. The medusae of some species of *Hebella* Allman, 1888, and *Anthohebella* gen. nov. (Cnidaria, Hydrozoa, Lafoeidae), with a world synopsis of species. *Zool. Verh.* **1997**, *310*, 1–53.

32. Bouillon, J.; Gravili, C.; Pagès, F.; Gili, J.M.; Boero, F. An introduction to Hydrozoa. *Mém. Mus. Nat. Hist. Nat.* **2020**, *194*, 1–591.

33. Sánchez, J.A.; Nava, G.R. Note on the distributions, habitat and morphology of *Stylaster roseus* (Pallas 1766) (Hydrozoa; Stylasterina) in the Colombian Caribbean. *Bol. Investig. Mar. Cost.* **1994**, *23*, 193–197.

34. Hoeksema, B.W.; van Moorsel, G.W.N.M. Stony corals of St. Eustatius. In *Marine Biodiversity Survey of St. Eustatius, Dutch Caribbean, 2015*; Hoeksema, B.W., Ed.; Naturalis Biodiversity Center: Leiden, The Netherlands, 2016; pp. 32–37.

35. Hoeksema, B.W.; Koh, E.G.L. Depauperation of the mushroom coral fauna (Fungiidae) of Singapore (1860s–2006) in changing reef conditions. *Raffles. Bull. Zool. Suppl.* **2009**, *22*, 91–101.

36. Reimer, J.D.; Wee, H.B.; García-Hernández, J.E.; Hoeksema, B.W. Zoantharia (Anthozoa: Hexacorallia) abundance and associations with Porifera and Hydrozoa across a depth gradient on the west coast of Curaçao. *Syst. Biodivers.* **2018**, *16*, 820–830. [CrossRef]

37. IUCN. *Coral Reef Resilience Assessment of the Bonaire National Marine Park, Netherlands Antilles*; International Union for Conservation of Nature: Gland, Switzerland, 2011; p. 51.

38. Parsons, G.; Thur, S. Valuing changes in the quality of coral reef ecosystems: A stated preference study of scuba diving in the Bonaire National Marine Park. Environ. *Resour. Econ.* **2008**, *40*, 593–608. [CrossRef]

39. Jackson, J.B.C.; Donovan, M.; Cramer, K.; Lam, V. (Eds.) *Status and Trends of Caribbean Coral Reefs: 1970–2012*; Global Coral Reef Monitoring Network, International Union for the Conservation of Nature Global Marine and Polar Program: Washington, DC, USA, 2014.

40. Bak, R.P.M. Ecological aspects of the distribution of reef corals in the Netherlands Antilles. *Bijdr. Dierk.* **1975**, *45*, 181–190. [CrossRef]

41. Ostarello, G.L. Natural history of the hydrocoral *Allopora californica* Verrill (1866). *Biol. Bull. Mar. Biol. Lab. Woods Hole* **1973**, *145*, 548–564. [CrossRef]

42. Boschma, H. The species problem in *Millepora*. *Zool. Verh.* **1948**, *1*, 1–115.

43. de Weerdt, W.H. Taxonomic characters in Caribbean *Millepora* species (Hydrozoa, Coelenterata). *Bijdr. Dierk.* **1984**, *54*, 243–262. [CrossRef]

44. Edmunds, P.J. The role of colony morphology and substratum inclination in the success of *Millepora alcicornis* on shallow coral reefs. *Coral Reefs* **1999**, *18*, 133–140. [CrossRef]

45. Sebens, K.P.; Witting, J.; Helmuth, B. Effects of water flow and branch spacing on particle capture by the reef coral *Madracis mirabilis* (Duchassaing and Michelotti). *J. Exp. Mar. Biol. Ecol.* **1997**, *211*, 1–28. [CrossRef]

46. Glynn, P.W. Increased survivorship in corals harboring crustacean symbionts. *Mar. Biol. Lett.* **1983**, *4*, 105–111.

47. Hoeksema, B.W.; van Beusekom, M.; ten Hove, H.A.; Ivanenko, V.N.; van der Meij, S.E.T.; van Moorsel, G.W.N.M. Helioseris cucullata as a host coral at St Eustatius, Dutch Caribbean. *Mar. Biodivers.* **2017**, *47*, 71–78. [CrossRef]

48. Montano, S.; Fattorini, S.; Parravicini, V.; Berumen, M.L.; Galli, P.; Maggioni, D.; Arrigoni, R.; Seveso, D.; Strona, G. Corals hosting symbiotic hydrozoans are less susceptible to predation and disease. *Proc. R. Soc. B* **2017**, *284*, 20172405. [CrossRef]

 *diversity*

Interesting Images

The Spotted Cleaner Shrimp, *Periclimenes yucatanicus* (Ives, 1891), on an Unusual Scleractinian Host

Ricardo González-Muñoz [1], Agustín Garese [1,*], Fabián H. Acuña [1], James D. Reimer [2,3] and Nuno Simões [4,5,6]

1   Instituto de Investigaciones Marinas y Costeras (IIMyC), Facultad de Ciencias Exactas y Naturales (FCEyN), Universidad Nacional de Mar del Plata (UNMdP)—Consejo Nacional de Investigaciones Científicas y Tecnológicas (CONICET), Laboratorio de Biología de Cnidarios (LABIC) (FCEyN), Rodríguez Peña 4046, Mar del Plata CC 1260. 7600, Argentina; ricordea.gonzalez@gmail.com (R.G.-M.); facuna@mdp.edu.ar (F.H.A.)

2   Molecular Invertebrate Systematics and Ecology Laboratory, Faculty of Science, University of the Ryukyus, Senbaru 1, Nishihara, Okinawa 903-0213, Japan; jreimer@sci.u-ryukyu.ac.jp

3   Tropical Biosphere Research Center, University of the Ryukyus, Senbaru 1, Nishihara, Okinawa 903-0213, Japan

4   Unidad Multidisciplinaria de Docencia e Investigación en Sisal, Facultad de Ciencias, UNAM, Puerto de Abrigo, Sisal, Hunucmá C.P. 97356, Yucatán, Mexico; ns@ciencias.unam.mx

5   International Chair of Coastal and Marine Studies in Mexico, Harte Research Institute, Texas A&M at Corpus Christi, TX 78412, USA

6   Laboratorio Nacional de Resiliencia Costera (LANRESC), Sisal, Hunucmá C.P. 97356, Yucatán, Mexico

*   Correspondence: agarese@mdp.edu.ar

Received: 30 October 2019; Accepted: 8 November 2019; Published: 12 November 2019

The spotted cleaner shrimp, *Periclimenes yucatanicus* (Ives, 1891), forms symbioses with sea anemones that may serve as cleaning stations for reef fishes [1]. This Caribbean palaemonid shrimp has usually been reported in symbiotic association with several species of actiniarian hosts, such as *Condylactis gigantea* (Weinland, 1860) and *Bartholomea annulata* (Le Sueur, 1817), or even with some corallimorpharians and a scyphozoan jellyfish [2]. During a field survey at Alacranes coral reef (26 June 2016; 22°27.14′ N, 89°45.79′ W; 13 m depth) on the Campeche Bank, Yucatán Peninsula, México, two spotted shrimps were observed swimming and walking above the polyps of the head coral *Montastraea cavernosa* (Linnaeus, 1767). Because none of the usual hosts of *P. yucatanicus* were detected nearby, we hypothesize that the shrimps were using the scleractinian coral as a host. Some other shrimp species commonly associated with actiniarians were previously reported to be living on stony corals, such as *Ancylomenes holthuisi* (Bruce, 1969) on *Heliofungia actiniformis* (Quoy and Gaimard, 1833) in New Guinea [3], and *Periclimenes rathbunae* Schmitt, 1924 on *Dendrogyra cylindrus* Ehrenberg, 1834 in Curaçao [4]. The observation (see Figure 1) of *Montastraea cavernosa* hosting *Periclimenes yucatanicus* is the second report of a palaemonid shrimp in association with a scleractinian coral in the Atlantic Ocean. The ecological implications of this association are unknown but could be related to a low local availability of usual hosts.

(a)   (b)

Figure 1. (**a**) Two anemone shrimps *Periclimenes yucatanicus* (arrows) on the scleractinian coral *Montastraea cavernosa*. (**b**) *P. yucatanicus* dorsal side. Scale bars: 1 cm.

Author Contributions: Conceptualization, R.G.-M., A.G., F.H.A.; investigation, A.G.; writing—original draft preparation, R.G.-M., A.G., F.H.A.; writing—review and editing, J.D.R.; project administration and funding acquisition, N.S.

Funding: This research was funded by grants to NS from the Harte Research Institute (Biodiversity of the Southern Gulf of Mexico) and CONABIO (NE018; Actualización del conocimiento de la diversidad de especies de invertebrados marinos bentónicos de aguas someras [<50 m] del sur del Golfo de México). We thank to José Luis Tello-Musi (UNAM) for his helpful assistance during field work.

Conflicts of Interest: The authors declare no conflicts of interest.

References

1. Titus, B.M.; Vondriska, C.; Daly, M. Comparative behavioral observations demonstrate the "cleaner" shrimp *Periclimenes yucatanicus* engages in true symbiotic cleaning interactions. *Royal Soc. Open Sci.* **2017**, *4*, 170078. [CrossRef]
2. Silbiger, N.J.; Childress, M.J. Interspecific variation in anemone shrimp distribution and host selection in the Florida Keys (USA): Implications for marine conservation. *Bull. Mar. Sci.* **2008**, *83*, 329–345.
3. Bruce, A.J. The hosts of the coral-associated Indo-West-Pacific Pontoniine shrimps. *Atoll Res. Bull.* **1977**, *205*, 1–19. [CrossRef]
4. Brinkmann, B.W.; Fransen, C.H.J.M. Identification of a new stony coral host for the anemone shrimp *Periclimenes rathbunae* Schmitt, 1924 with notes on the host-use pattern. *Contrib. Zool.* **2016**, *85*, 437–456. [CrossRef]

MDPI

St. Alban-Anlage 66

4052 Basel

Switzerland

Tel. +41 61 683 77 34

Fax +41 61 302 89 18

www.mdpi.com

Diversity Editorial Office

E-mail: diversity@mdpi.com

www.mdpi.com/journal/diversity